A
Simplified Guide to
Custom Stairbuilding
and
Tangent Handrailing

A
Simplified Guide to
Custom Stairbuilding
and
Tangent Handrailing

by George R. di Cristina

Linden Publishing, Inc.

Library of Congress Cataloging-in-Publication Data

Di Cristina, George R.
    A simplified guide to custom stairbuilding and tangent handrailing
    by George R. di Cristina.
    288 p. 21.6 x 27.6 cm.
    Includes index.
    ISBN 0-941936-27-9
    1. Stair building. 2. Hand-railing.   3. Wooden stairs.
TH5670.D5   1994   94-7753
694'.6--dc20   CIP

A Simplified Guide to Custom Stairbuilding and Tangent Handrailing
by George R. di Cristina
Copyright © 1994 by George R. di Cristina

135792468
Linden Publishing Inc.
3845 N. Blackstone
Fresno, CA USA 93726
Phone 1-800-345-4447

The author and publisher wish to thank Mr. Terry Hale of Quality Stairs, Fresno, CA, and Mr. Homer Eaton of J. di Cristina & Son, San Francisco, for their help in the preparation of this book.

Printed in United States of America

Contents

List of Plates

List of Photographs

Elegant custom-made residential stair with the starting four treads curved. All treads are 1⅛" thick of clear red oak. Risers and stringers are vertical-grain Douglas Fir. The wall stringer is a 1" thick housed and wedged type. The 1¼" thick laminated rail-side stringer is mitered to the ¾" thick risers. All risers extend to the bottom of the stringers to form a natural warped soffit line beneath the stair, to receive plaster. This serves as the ceiling line above a closet beneath the stair. A 3" x 4" selected-grain Honduras mahogany handrail is made by the tangent principle. The rail, easing from the starting level volute cap, gracefully sweeps over and around the mitered stringer and a level "S" curved balcony to terminate against the wall. 1½" balusters are turned from clear Douglas Fir.

Foreword

Although many of the stairbuilding practices of the past century and before are still performed in the same manner today, much of the extravagantly detailed work in both design and joinery has either been modified or disregarded altogether. Modern design, along with more efficient woodworking equipment and stronger and more enduring wood adhesives, has served to reduce both labor and material costs without sacrificing sound construction and aesthetic quality.

In order to obtain the most desirable effect in planning a staircase having a continuous inclined and side-curved handrailing, it is essential to have basic knowledge of the principles of tangent handrailing. Certainly, it would be to the stairbuilder's benefit to know tangent handrailing in its entirety. However, minimal knowledge is all that is required in order to properly place the risers in the stair body to achieve the most aesthetic appearance possible.

Although this book deals with properly building stairs, both straight and curved, the primary focus is how to make continuous climbing-turn handrail sections using the tangent principle. The tangent principle is a method for using tangent lines to make exact joints and patterns for any incline-turn handrail section.

During the early years of my 48 years of working in tangent handrailing, I briefly studied and applied the works of several turn-of-the-century authors. It soon occurred to me that there should be a less complex and more compact treatise on the subject. I then developed the idea of making small wooden blocks and folding cardboard models of the blocks, both simulating every possible tangent handrail condition. The resulting study of the blocks and folding models became the foundation for the compact, simplified, and unerring method of tangent handrailing I have successfully practiced for most of my career. It is this same method of tangent handrailing I now present. I believe that any dedicated student will readily understand the tangent method of handrailing I have shown in this book, and will be able to practice it with confidence.

I dedicate this book to the treasured memories of my father John and my brother Charles, founders of J. di Cristina & Son. They were my inspiration. Most of all, I dedicate it to my wife, whose understanding and encouragement were unflagging during the long process of writing and detailing the manuscript.

Introduction

A Simplified Guide to Custom Stairbuilding and Tangent Handrailing

The purpose of this book is threefold. The first is to present to the wood-working craftsman the method of making both straight and curved stairs in an aesthetic and sound stairbuilder's fashion. The second is to provide a step-by-step guide for making incline-turn handrail sections through a precise tangent principle. Finally, my purpose is to teach the use of the tangent principle for designing stairs with continuous incline-turn handrail sections.

The text covers the construction method of straight-line stairs where there are no curves in the supporting tread and riser members (called the "stringers"), and for those curved or circular stairs where the stringers may or may not be completely curved.

Straight-line stairs may include a straight run of treads, platform turns or winding treads. They may also include curved and continuous handrailings. The stringers of such stairs may be recessed to receive the treads and risers, cleated to receive treads, mitered at one or both sides of the stair to receive mitered risers, or be cut out to receive treads and risers. A stringer may also be any supporting member.

Curved stair stringers are either laminated, staved, or kerfed-and-keyed. Lamination is the gluing and layering of many thin members to produce the stringer thickness. A staved stringer is one in which many narrow solid pieces, beveled at both sides of the width, are glued side by side to a thin veneer. A kerfed-and-keyed stringer is a solid piece of stringer stock closely kerfed at the back side so that the stringer can be bent around a curved form where tapered wooden keys are inserted and glued into the saw kerf. Staved-and-keyed stringers are generally used for tight-radius conditions. A plate covering the method of staving and kerfing is shown in this book. Owing to the excellence of present-day wood adhesives, the practice of laminating incline-turn handrailings is used extensively, especially in stock circular staircases, or in custom-made stairs involving turns where the radius is great enough to allow for the bending. However, laminating a handrail is limited to certain conditions and may often require adjustment. The tangent principle, on the other hand, encompasses all possible handrail conditions regardless of the radius or pitch, with no adjustments necessary.

Stairbuilding and tangent handrailing will be shown in five sections:

Section I is confined to stairs having only straight-line construction in the stair body, but may or may not involve a continuous handrailing around a turn.

Section II is concerned with curved stair planning and construction involving curved and inclined handrail sections.

Section III deals with making curved handrail sections for curved staircases or any other such climbing-turn conditions.

Section IV shows a short, or abbreviated form of finding the face mold pattern essential for making a climbing-turn handrail section.

Section V shows how to make handrail sections to fit on top of existing metal railing through the tangent principle.

Prior to getting involved with the above sections, it is essential to first show some of the basic geometrical figures associated with both stairbuilding and tangent handrailing. These are shown in **Plates 1, 2,** and **3.**

Plate 1—Basic Geometrical Figures Useful in Stairbuilding and Tangent Handrailing

Figures 1, 2, and 3 show the three basic angles. **Figure 1** shows a 90 degree right angle; Figure 2, an obtuse angle (greater than 90 degrees); and **Figure** 3, an acute angle (less than 90 degrees). All angles intersect the radius curves at a and c. oc and oa are perpendicular to the legs of the angles.

Figure 4 shows parallel lines which are equal distance apart throughout their lengths. **Figure 5** is a square showing all sides equal length, and all angles being 90 degrees. **Figure 6** is a rectangle where opposite sides are equal length, and all angles are also 90 degrees. **Figure 7** shows an oblique parallelogram where opposite sides are equal length, parallel and slanted. Opposite angles are equal.

Figure 8, 9, and **10** show methods of constructing a perpendicular on, or at the end of a line.

Figure 8—To draw a perpendicular line from point c, make equal lengths cd at each side of c. With d as radius centers, strike arcs of equal length to intersect at e. Connect c to e for the required perpendicular.

Figure 9—To draw a perpendicular at the end of a line, say at b along line ab, with the compass point well above line ab, say at c, and with radius cb, strike a wide arc to intersect ab at d. With a straight-edge, draw a line from d through c to intersect the arc at e. Connect b to e for the required perpendicular.

Figure 10 is the more common practice of drawing a perpendicular, or a right angle, at the end of a line. To draw a perpendicular from point b, let d equal 3 units; e, 4 units; and f, 5 units, or d, e, and f, may be multiples of 3, 4, and 5.

Plate

1

Basic Geometrical Figures Useful in Stairbuilding
and Tangent Handrailing

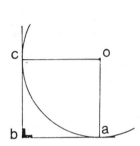

Figure 1

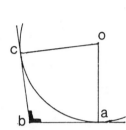

Figure 2

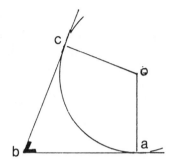

Figure 3

Figure 4

Figure 5

Figure 6

Figure 7

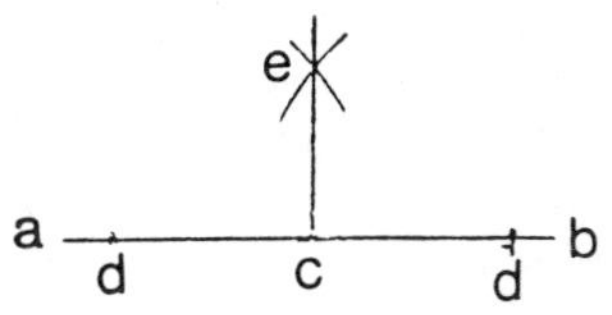

Figure 8

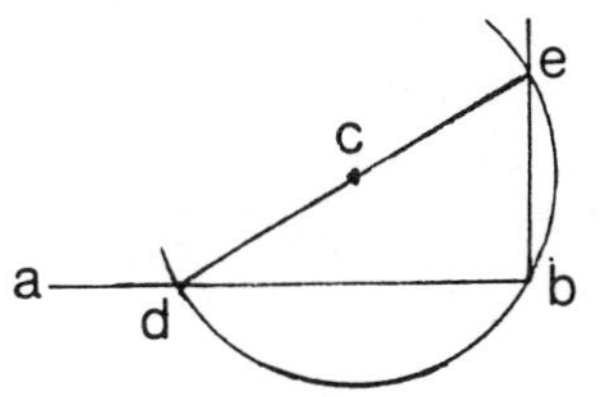

Figure 9

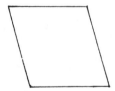

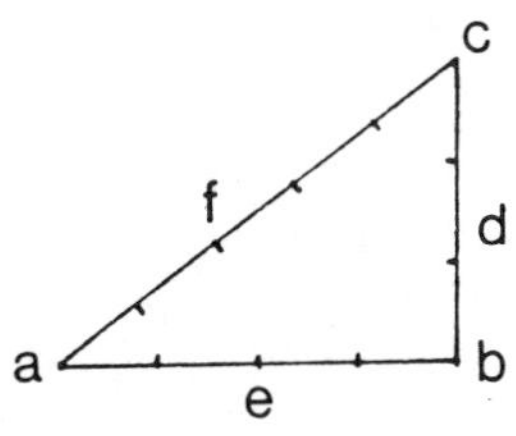

Figure 10

<table>
<tr><td>

Plate

2
</td><td>

Three Methods of Drawing a Semi-Ellipse
</td></tr>
</table>

When we come to the handrail section of this book, showing how to make the pattern known as the face mold for making a climbing-turn handrail section, the curves of the face mold will be found to be a portion of a semi-ellipse whenever the plan curves are drawn with a radius.

An example of a semi-ellipse is easily understood in **Figures 1** and **2.** Assume **Figure 1** represents the end view of a solid half-round length of material. OC is the radius, and AB is the diameter. If the half-round length is cut across its diameter at the pitch A'B' of **Figure 2**, the shape of the curve at the oblique cut will no longer be the curve of a radius-drawn semi-circle, but will be the curve of a semi-ellipse. A'B' is termed the long diameter, or major axis. O'C' is termed the short diameter, or minor axis. This plate shows three methods of drawing a semiellipse. **Figure 2** is the ordinate method; **Figure 3**, the trammel method; and **Figure 4**, the string-and-pin method.

Figure 2—The ordinate method: 1', 2', 3', 4' and 5' are the lengths from A'B' to the curve, the same lengths as in **Figure 1** from AB to the plan curve.

Figure 3—The trammel method: On a rod longer than A'O', mark 1-2 equal to O'C'. 2-3, marked on the same rod, equals O'A'. Mark random points at 2 while keeping mark 1 along the major axis and mark 3 along the extended line of the minor axis O'C'. Connect all points thus marked at 2 for the curve.

Figure 4—The string-and-pin method: With A'B' the same pitch, and O'C' the same length as in **Figures 2** and **3**, mark C'<u>a</u> and C'<u>b</u> equal to A'O'. Mark <u>a</u> and <u>b</u> as pin

placements. With a linen line fastened to either pin, <u>a</u> or <u>b</u>, loop the line around the other pin several times. With a notched pencil point, stretch the line to C'. Holding the line taut at C', draw the semi-ellipse from A' to B' as shown at random locations E. <u>ef</u> will be explained in another plate.

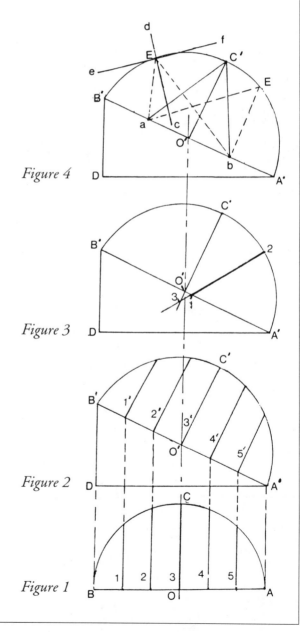

Figure 4

Figure 3

Figure 2

Figure 1

Plate 3

Right, Obtuse, and Acute Angles Formed by Tangents in The Plans of Handrail Sections

Figure 1 is an example of the tangent line. The tangent line is a line touching a curve., perpendicular to the curve, whether or not the curve is drawn from a radius point. In **Figure 4, Plate 2**, <u>ef</u> is tangent to the elliptical curve at point E. It is perpendicular to <u>cd</u>, the bisector of angle <u>aEb</u>.

The three basic angles of **Plate 1** are shown here as tangents of handrail sections. **Figure 2** shows a right angle plan, **Figure 3** is an acute angle plan, and **Figure 4** is an obtuse angle plan. Equal length tangents are drawn from spring lines S at the centerlines of the handrail sections to form the three angles. The tangents are always perpendicular to the radius, or spring lines.

Tangent line

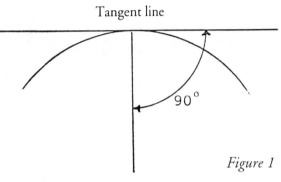

Figure 1

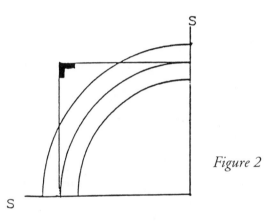

Figure 2

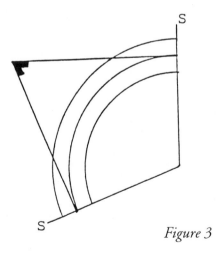

Figure 3

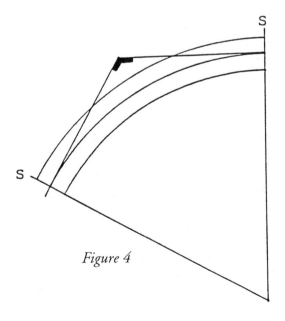

Figure 4

Section I

Straight Stairs

Stairbuilding begins with the stair plan. The area allowed for the stair is referred to as the stairwell. Some stairwells limit the choice of stair plans, while others are spacious enough for a choice of several plans.

There are a number of types of stair plans: straight, platform, winder, and curved or circular. There may, or may not, be posts, balusters (upright supports under a handrail), or straight or curved, continuous handrailings. Whatever the choice, the stair should suit the space allowed as well as the building's architecture.

The first step before determining the proper stair plan to fit within the dimensions of the stairwell is to calculate the number of risers contained in the floor-to-floor height. The "rise" is the riser height between two successive treads. The tread "run" or "cut" is the horizontal distance between the two risers of a tread.

While some stairbuilders may use specific formulas for determining tread runs to riser heights, I have found the following table of tread runs and riser heights a satisfactory guide for a comfortable ascending and descending stair. The table shown should be considered as a guide and adhered to whenever possible.

In a public or commercial main stair, the tread run should not be less than 10", and the rise not greater than 7-1/4". A 10" run is a good standard to follow in a residential main staircase, with a riser height between 6-3/4" to 7-5/8". Of course, there are conditions when these desired dimensions cannot always be followed.

A Guide to Tread and Riser Proportions

When the run is:	The rise should be:	Stair use:
9"	7-⅞" to 8-½"	Utility
9-½" to 9-¾"	7" to 7-¾"	Utility - Residential
10"	6-¾" to 7-¼"	Public - Residential
10"	7-½" to 7-¾"	Residential
11"	6" to 7"	Public - Residential
12"	5" to 6-½"	Public
14"	4-½" to 5-½"	Public
16" to 18"	4-½" to 5"	Public
20" and wider	4"	Public

Once the number of risers for a stair has been determined, the perimeter of the stairwell should be drawn to a scale of ¾" or 1" equal to 1', and the stair plan drawn within to suit.

Details of treads, risers, and stringers are shown in **Plates** 4 through 11 followed by types of stairs involving straight runs, platforms, and winders. These stairs do not have curved stringers but may have a curved continuous handrailing such as in **Plate 13**, showing a platform turn with straight stair runs at both sides of the turn.

Handrail volutes and bull-nosed-end treads are shown in **Plates 14** through **15**, with the method of constructing a bull-nosed tread in **Plate 16**. Subsequent plates up to **Plate 27** show details of stringers, posts, and handrail conditions that might occur for the straight-line the plan types of **Plate 12**. This will conclude the straight-line stair construction shown in this section.

Plate 4—The Tread and Riser in Elevation

Figure 1 shows typical tread and riser connections. The risers may be simply glued and nailed to the back of the tread, and the tread glued and nailed to the top of the riser. The bottom of the tread may be grooved to receive the riser, and the back of the riser may be grooved to receive the back end of the rabbeted tread. Treads with the same width throughout their lengths are called "square" or "straight" treads. The distance between risers is called the "run" or "cut". As mentioned earlier, the riser height between two treads is often referred to as the "rise".

Figure 2 illustrates the importance of a nosing projection, or overhang, being forward of the riser. With a projection, the heel of a descending shoe will not likely contact the face of the riser, such as at a. However, without the nosing projection of the tread, there is a possibility of the heel contacting the tread at b, causing the person to pitch forward.

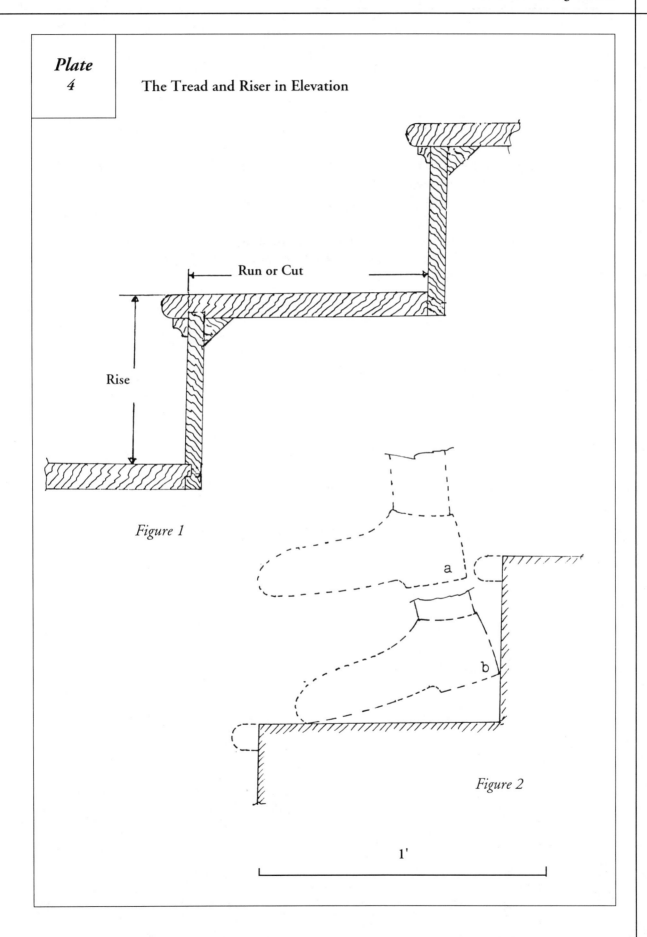

Plate 4

The Tread and Riser in Elevation

Run or Cut

Rise

Figure 1

a

b

Figure 2

1'

Plate 5—Kick-Back Riser and Tread in Elevation and Landing Details

Figure 1 elevation shows the riser beveled back from the nosing of the tread as a "kick-back" riser. It is an alternative to the nosing projection shown in **Plate** 4. The tread and riser may also be rabbeted or grooved as in **Plate** 4.

Figures 2 and **3** show typical landing details at the headers of the platform or stair landing levels. The "landing strips" are rabbeted in various ways to suit the floor conditions.

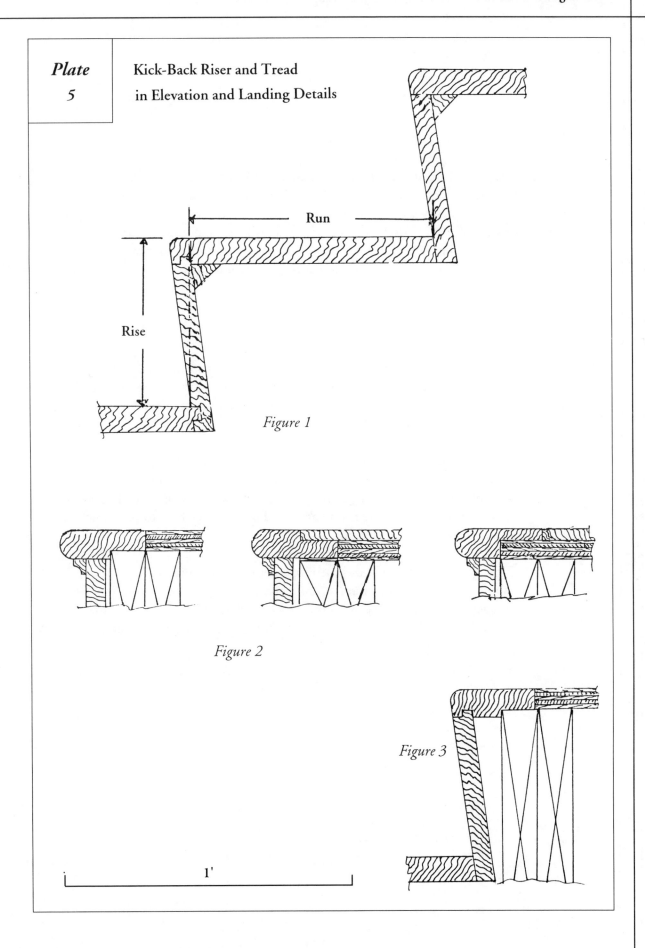

Plate 5 — Kick-Back Riser and Tread in Elevation and Landing Details

Run

Rise

Figure 1

Figure 2

Figure 3

1'

Plate 6—The Stringer, Pitch Board, and T-Gauge

Although there are many types of staircases to suit conditions and one's imagination, common to all is obviously the succession of elevated steps to get from one level to another. Other than in an unusual modernistic design of stairs where treads may project from a wall, be supported by central beams, hang from supports from the ceiling, or rest on supports from below, the traditional solid and continuous supporting members of treads and risers are called "stringers".

The stringer – There are various types of traditional stringers: carriage, housed and wedged, mitered-to-riser, mitered-to-bracket, buttress, dapped to receive treads only, and cleated to receive treads. They are detailed in the following plates.

The pitch board – In order to quicly mark out the treads and risers on a stringer, a template called the pitch board is best used. It is a pattern of the pitch of the hypotenuse of a right triangle formed by the riser height and the tread. It is generally made from hardboard or plywood 1/8" to 3/8" thick. It is used in conjunction with the T-gauge.

T-gauge – The pitch board is used together with a margin guide called the T-gauge. The T-gauge maintains a uniform margin from the bottom of a stringer to the meeting point of the tread and riser. This plate shows the use of the pitch board and T-gauge for laying out a carriage type stringer where pieces are cut out.

Plate 6

The Stringer, Pitch Board, and T-Gauge

Pitch board

Cut out

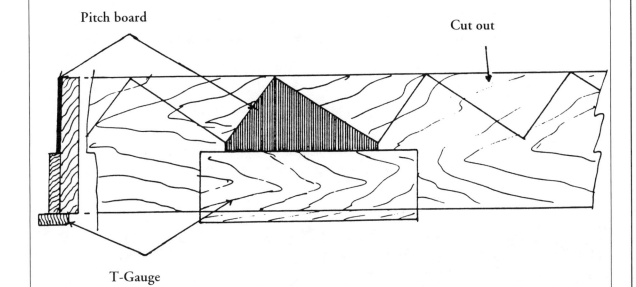

T-Gauge

1'

Plate 7—The Stringer and Timber

Plate 7 is a housed and wedged type stringer. Its thickness should be no less than ¾".

In **Figure 1,** the stringer is shown receiving both the treads and the risers which are wedged and glued into the housing. The housing must be tapered to receive the tapered wedges.

Figure 2 is a section of **Figure 1** showing the stringer at a wall. This stringer is approximately 1" thick and rabbeted at the top to receive ½" wallboard, leaving a minimum of ½" of stringer to receive baseboard at the bottom and top of the stringer. The depth of the housing to receive the treads, risers, and wedges is ½".

Figure 3 shows the cut at both top and bottom to receive the baseboard. It also shows the "timber". The timber is either a kiln-dried 2x4 or 2x6 piece of common stock supporting member between stringers. At alternating sides of the timber, either 1x8 kiln-dried Douglas Fir or ¾" plywood brackets are glued and nailed. They are placed tight against the tread and riser and well glue-blocked.

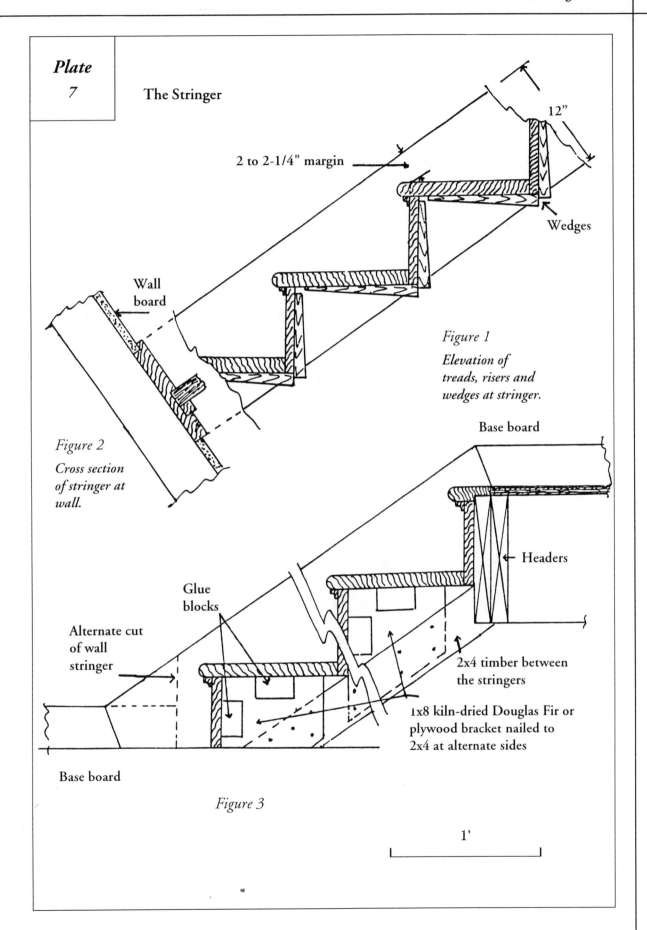

Plate 7

The Stringer

12"

2 to 2-1/4" margin

Wedges

Wall board

Figure 1

Elevation of treads, risers and wedges at stringer.

Figure 2

Cross section of stringer at wall.

Base board

Headers

Glue blocks

2x4 timber between the stringers

Alternate cut of wall stringer

1x8 kiln-dried Douglas Fir or plywood bracket nailed to 2x4 at alternate sides

Base board

Figure 3

1'

Plate 8—A Mitered Face-Type Stringer

Figure 1 is the elevation of a 1" thick stringer. It is mitered to receive mitered end risers.

Figure 2 is a section of **Figure 1** assuming a wall beneath the stringer. The horizontal 2x4 is the pitched plate on top of the studs. The edged 2x4 is a supporting member under the stair at the treads and risers. The 1" thick mitered face stringer is shown rabbeted at the bottom to receive ½" wallboard.

Figure 3 is a top view of the riser mitered to the stringer. The stringer is glueblocked to both treads and risers.

Figure 4 shows a top view of the returned end of a typical tread. The tread shows a nosing return piece to project from the back of the tread as much as the front nosing of the tread projects forward of the riser. See **Figure 1**.

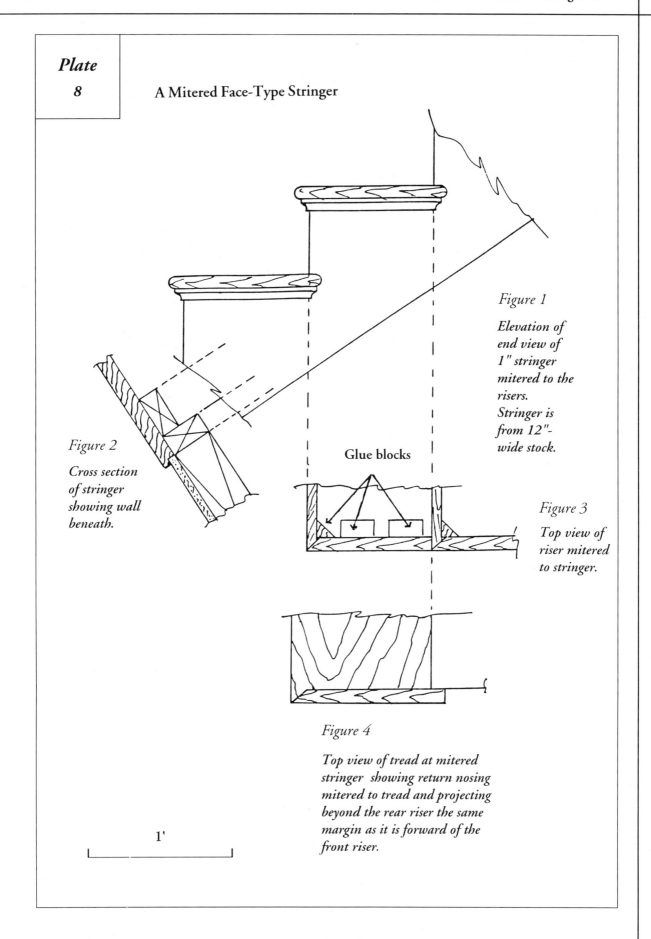

Plate 8

A Mitered Face-Type Stringer

Figure 1

Elevation of end view of 1" stringer mitered to the risers. Stringer is from 12"-wide stock.

Figure 2

Cross section of stringer showing wall beneath.

Glue blocks

Figure 3

Top view of riser mitered to stringer.

Figure 4

Top view of tread at mitered stringer showing return nosing mitered to tread and projecting beyond the rear riser the same margin as it is forward of the front riser.

1'

Plate 9—A Mitered Bracket-Type Stringer

Figure 1 is an elevation showing the face of the stringer with the bracket applied to the stringer and mitered to the riser. The bracket shown here is ⅜" thick.

Figure 2 is a top view showing the riser mitered to the 3/8" bracket and nailed and glued to the stringer.

Figure 3 shows a top view of the tread with a special return nosing much like the return nosing shown in **Plate 8**, **Figure** 4, except that the nosing is ⅜" thicker and longer at the back of the tread to receive the bottom of the bracket. See **Figure 1**.

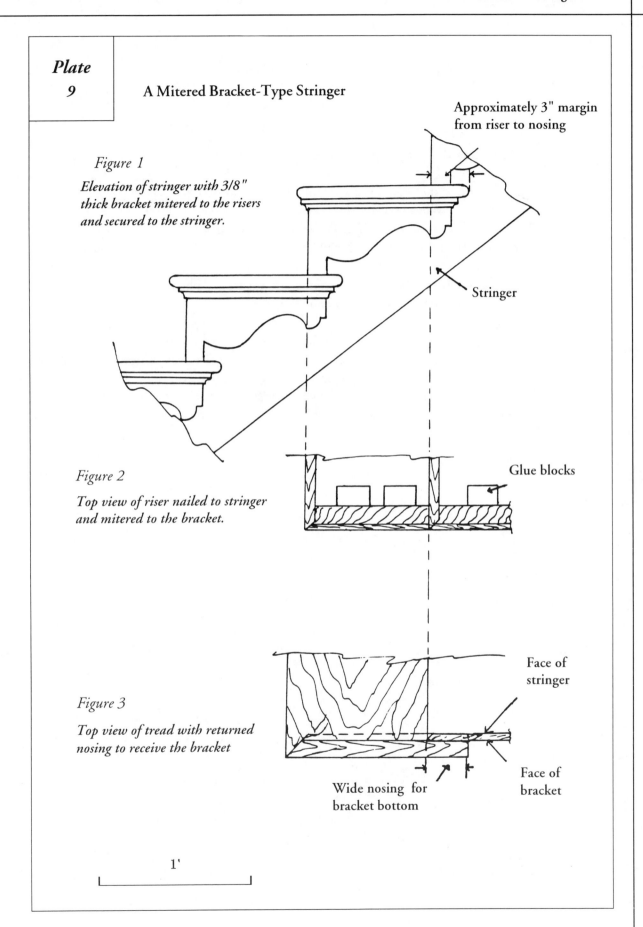

Plate 9

A Mitered Bracket-Type Stringer

Approximately 3" margin from riser to nosing

Figure 1

Elevation of stringer with 3/8" thick bracket mitered to the risers and secured to the stringer.

Stringer

Figure 2

Top view of riser nailed to stringer and mitered to the bracket.

Glue blocks

Figure 3

Top view of tread with returned nosing to receive the bracket

Face of stringer

Face of bracket

Wide nosing for bracket bottom

1'

**Plate 10—A Housed and Wedged Buttress-Type Stringer
and a Recessed, or Dapped, Stringer**

Figure 1 shows the typical elevation of a housed and wedged type stringer. Since there is to be a baluster shoe on top of the stringer in order to receive balusters for a handrail, the buttress width must be arbitrarily 3" to 3-½" wide. The buttress width is to include a fascia.

Figure 2 shows the buttress width to include the fascia. The fascia is shown rabbeted at the bottom to receive wallboard.

Figure 3 shows a 1-½" minimum thickness by nominal 12" width solid stringer. It is recessed, or "dapped," approximately ¾" to receive both treads and risers or just treads. In this case it is to receive treads and risers. If balusters are required, the stringer may have to be 2-½" thick or thicker. If the stringer is 1-½" thick, balusters can be secured to the side.

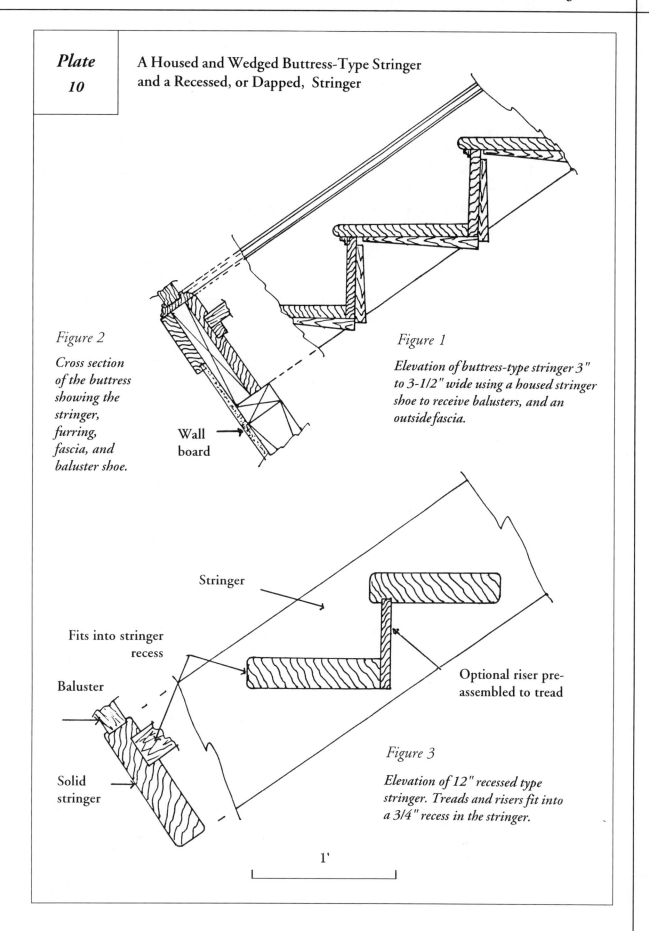

Plate 10

A Housed and Wedged Buttress-Type Stringer and a Recessed, or Dapped, Stringer

Figure 2

Cross section of the buttress showing the stringer, furring, fascia, and baluster shoe.

Wall board

Figure 1

Elevation of buttress-type stringer 3" to 3-1/2" wide using a housed stringer shoe to receive balusters, and an outside fascia.

Stringer

Fits into stringer recess

Baluster

Optional riser pre-assembled to tread

Figure 3

Elevation of 12" recessed type stringer. Treads and risers fit into a 3/4" recess in the stringer.

Solid stringer

1'

Plate 11—Two cleated type stringers

Figure 1 is an elevation of an interior only wood-cleated, 1-½" minimum thickness, solid, 12" wide stringer to receive 1-½" to 2-½" thick solid treads. The cleats are glued and screwed to both the stringer and treads.

Figure 2 is the cross section showing the treads butted to the stringer with the cleats glued and screwed.

Figure 3 is an exterior example similar to **Figure 2** except that there are stock galvanized metal tread angles used instead of wooden cleats. Prior to installing, the metal angles should be well primed with rust-resisting paint. This elevation shows 2-½" thick, two-piece treads. The angles are lag-bolted to the treads and either lag-bolted or machine-bolted to the stringers.

Figure 4 cross section shows the tread to be approximately 3/16" clear of the stringer, and the stringer bolt to receive two 1-½" diameter galvanized washers between the angle and the stringer so as to allow for drainage.

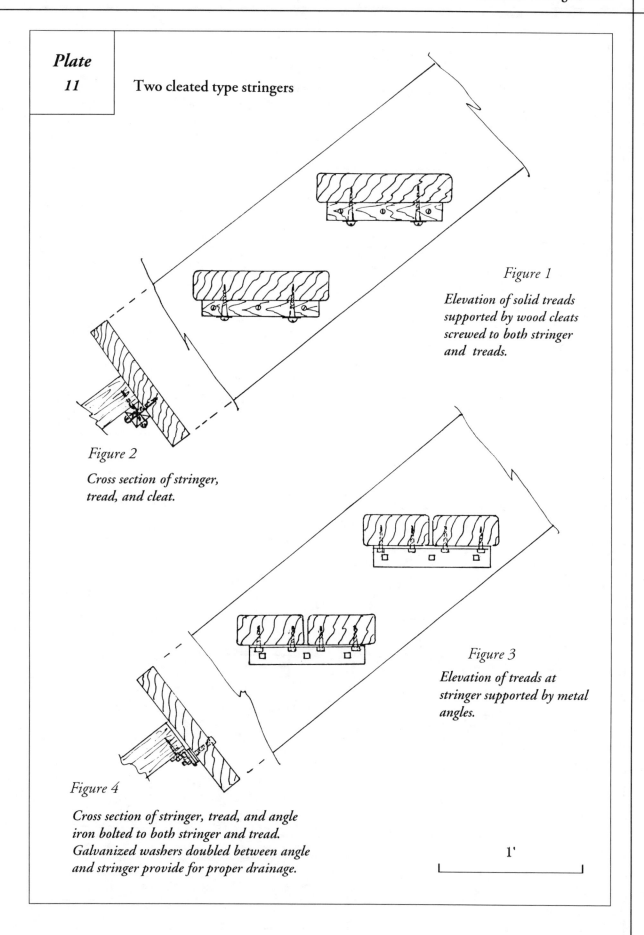

Plate 11   Two cleated type stringers

Figure 1

Elevation of solid treads supported by wood cleats screwed to both stringer and treads.

Figure 2

Cross section of stringer, tread, and cleat.

Figure 3

Elevation of treads at stringer supported by metal angles.

Figure 4

Cross section of stringer, tread, and angle iron bolted to both stringer and tread. Galvanized washers doubled between angle and stringer provide for proper drainage.

1'

Plate 12

Types of Stairs Involving Straight Runs, Platforms, and Winder Treads

Plans A through R may be between walls or open at one or both sides using any of the preceding stringer types. All plans have 15 risers except Plan B. The starting riser of each stair begins at the arrow. Some plans may have continuous handrails.

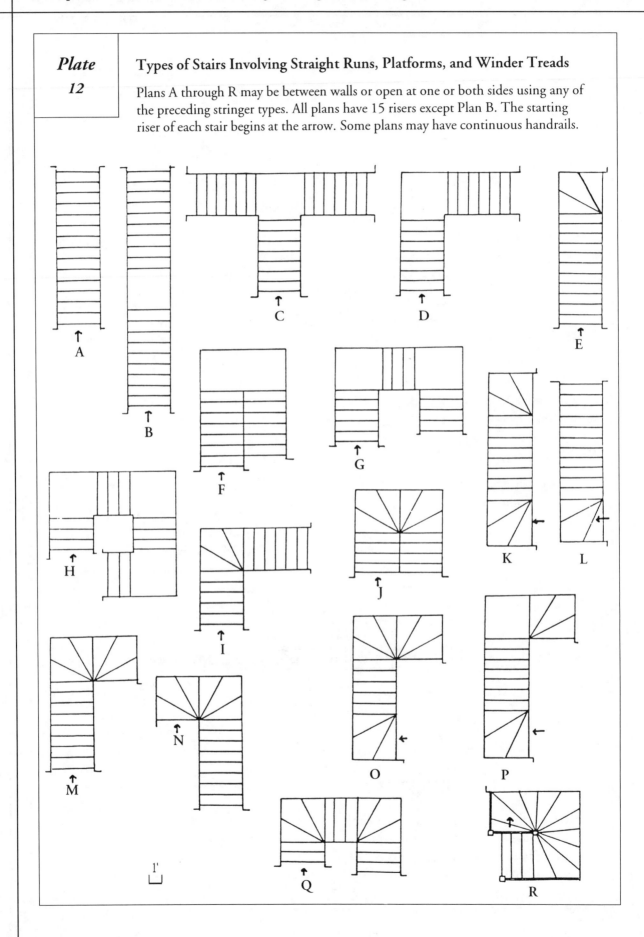

Plate 13

Riser Positions at the Right Angle Corner of a Platform Stair

To receive a continuous handrailing around the inside corner of a platform-type stair, the distance between landing and starting risers at the platform of two straight runs of treads should be properly established so that the turn section of handrail will meet the pitch of the rail at the straight runs without the necessity of additional easements. This is accomplished whenever the centerline distance, including tangents and straight rail from the landing to starting platform risers, is equal to the width of a normal square tread. The continuous handrail can be made to suit any placement of the plan risers. However, the most satisfactory results, using minimal material and labor, are obtained whenever the risers are located as mentioned.

Figure 1 is the plan of such a platform-type stair. AB plus BC plus d equals the 10" dimension between landing riser 2 and starting riser 3, that of a normal square tread. **Figure 2** shows the stretch out of the plan rail centerline, plan tangents, the risers in elevation, and the centerline pitch of the rail and tangents touching all tread and riser points. The pitched tangents are indicated by the letter t. It should be noted here that the plan and pitched tangents in this, and all handrail layouts throughout this book, do not represent the stretch out of the centerline curves of the handrail. Tangents are simply the instruments through which the incline-turn handrail section is made. The making of such a rail will be shown in **Section III**.

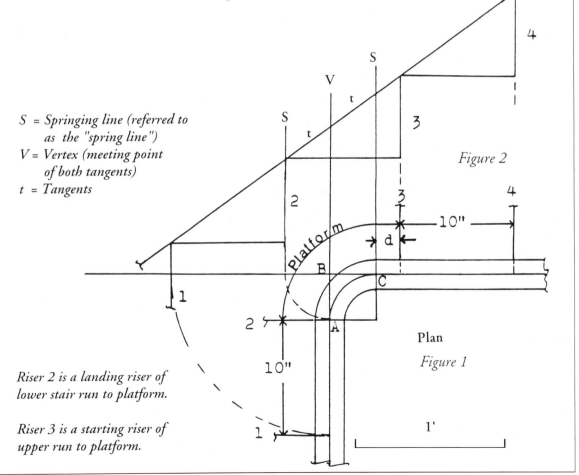

S = Springing line (referred to
    as the "spring line")
V = Vertex (meeting point
    of both tangents)
t = Tangents

Figure 2

Figure 1

Plan

Riser 2 is a landing riser of lower stair run to platform.

Riser 3 is a starting riser of upper run to platform.

Plate 14—The Handrail Volute and Turnout for Starting Bull-nosed Treads

Of several methods of drawing a volute or turnout at the start of a stair, I have found the following method simple and very satisfactory. Rather than terminate the scroll to a sharp vee at the rail cap, I have chosen to allow for finger room clearance.

In **Figures 1** through 4, let ab equal the scroll width, c equal the width center, and cd equal 1". Draw the outside radius quarter circle with ad as the radius. Let ef equal db, and draw a quarter circle with radius ef. Let eg equal 1". In **Figures 1** through **3**, draw quarter circles with g and h as radius points to make the rail-cap to suit. In **Figure 4**, let d be a radius point to complete the rail-cap.

To draw the turnout in **Figure 5**:

1 - Draw spring line ab.

2 - Let ab be an arbitrary radius.

3 - Draw arc bc to suit.

4 - At any point on ac draw radius dc.

5 - From d, draw a perpendicular to ab to intersect radius dc at e.

6 - At any point along de, such as at f, draw the level cap as arc eh to meet the normal rail width as shown.

Plate 14   The Handrail Volute and Turnout for Starting Bull-nosed Treads

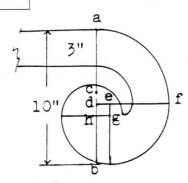

Figure 1

Closed scroll

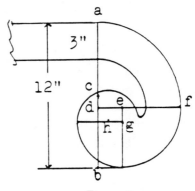

Figure 2

Open scroll

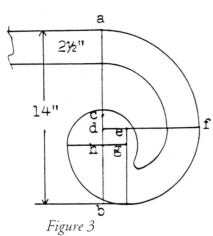

Figure 3

Open scroll

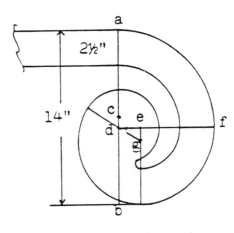

Figure 4

Closed scroll

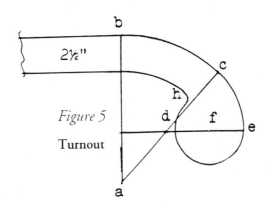

Figure 5

Turnout

1'

Plate 15—Bull-Nosed Treads for Stairs with an Open Side

Figures 1 through 4 show types of bull-nosed starting treads to receive either a starting post, handrail turnout, or a starting volute. Certain bull-nosed treads are available from stock stair suppliers but may be restricted to wood types and dimensions. However, these treads can be custom made to suit. See **Plate 16.**

Figure 5 shows the method of determining solid wood between kerfs of a bull-nosed tread riser to be bent to a 5-½" radius (also see **Plate 16**).

In **Figure 5**, take a piece of wood the same thickness as the riser to be bent, approximately 18" long and about 1" wide.

Approximately 8" from one end, make a saw kerf using the same saw that will be used to make the kerfs on the riser, leaving a maximum of 1/16" of wood remaining. On a board, draw a straight line at least 18" long as x-x in **Figure 5**. Lay the kerfed edge of the piece along the straight line and mark the kerf point as r. At r strike a 5-½"-radius arc. Secure the 10" end of the piece to the board with brads as shown. Now, bend the 8" portion until the kerf just closes, noting the distance the piece has moved along the arc as ab. ab will then be the solid wood required between kerfs of the riser. Stairbuilding author Robert Riddell introduced this practice in his treatise published at the turn of the nineteenth century.

Plate 15

Bull-Nosed Treads for Stairs with an Open Side

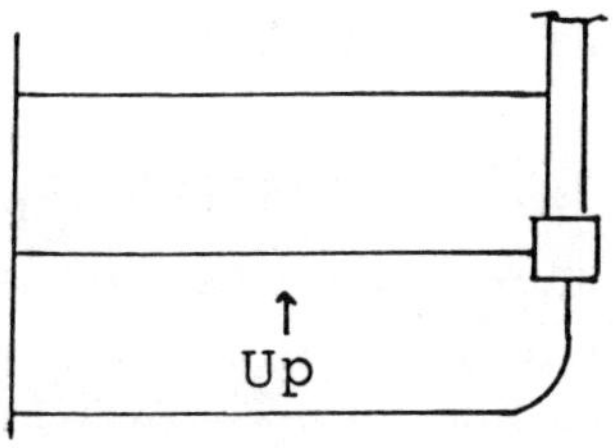

Figure 1

Post at start

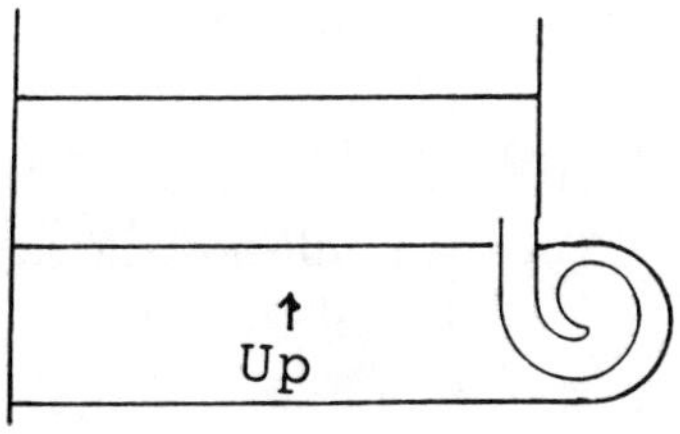

Figure 2

Small closed volute

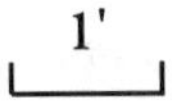

1'

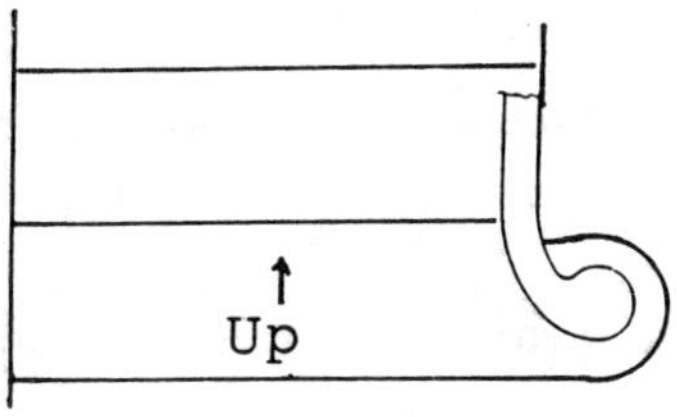

Figure 3

Turnout

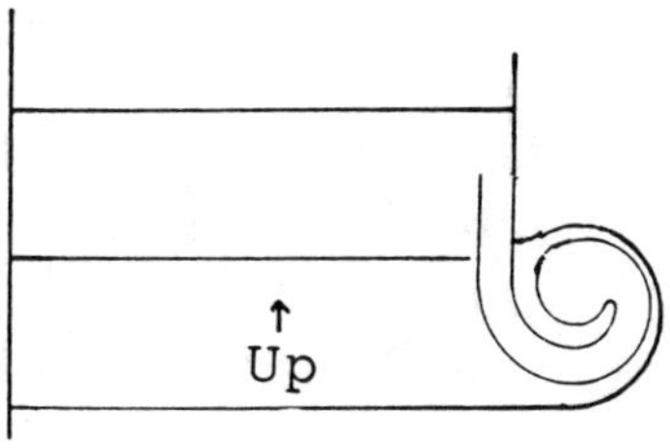

Figure 4

Tread for large volute

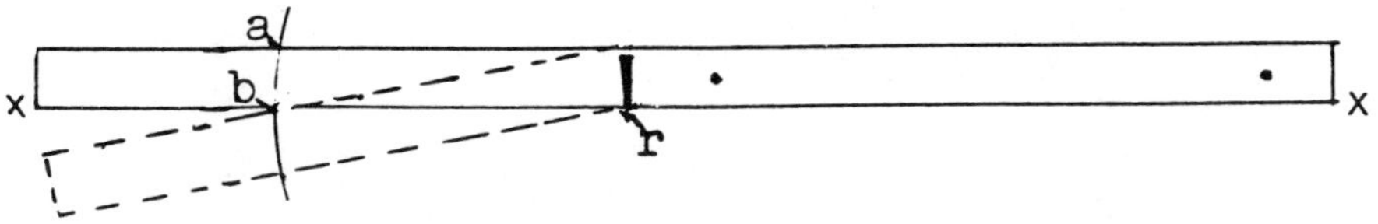

Figure 5

r = radius point

Plate 16—Bending a 5-½" Radius Riser for a Bull-Nosed Tread

A ¾"-thick riser, say 5-½" wide, is to be bent to a 5-½" radius around the core of a bull-nosed tread as shown in **Figure 1**. The length of riser to be kerfed is taken with a string or thin strip. Ideally, if the kerfing is done as shown in **Plate 15**, the kerf marks will not show at the riser face. If kerfs are not spaced in the manner shown, then a good practice to prevent the kerf marks from showing along the riser face is to first veneer approximately 1/16" of the riser face several inches past the length of the point where the kerfing will end. Then kerf the remaining solid portion of the veneered end of the riser leaving approximately 1/16" of wood remaining. In order to prevent the riser from possibly breaking during the bending, it is also a good practice to use a thin sheet-metal backing piece.

Figure 1 is a plan view of the tread bottom side up showing the form cores for bending the riser.

Figure 2 shows a side view of the tread bottom side up with a core secured to the tread and a core glued to the riser as the riser is bent.

Figure 3 is an edge view showing the 1/16" veneer x, the sheet metal backing piece, and the solid wood y to be kerfed.

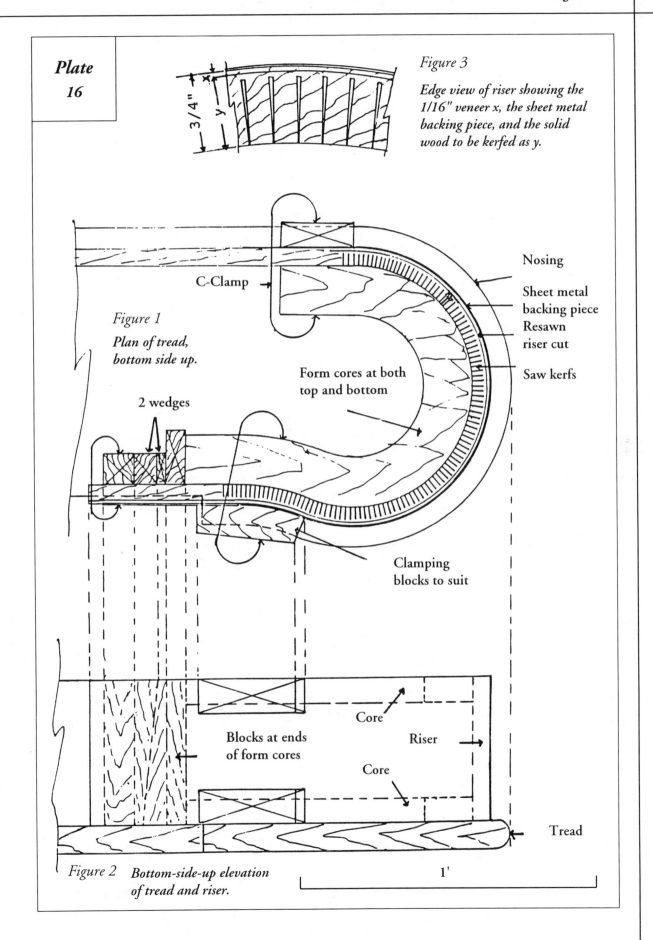

Plate 16

Figure 3

Edge view of riser showing the 1/16" veneer x, the sheet metal backing piece, and the solid wood to be kerfed as y.

3/4"

x

y

C-Clamp

Nosing

Sheet metal backing piece

Resawn riser cut

Saw kerfs

Figure 1

Plan of tread, bottom side up.

Form cores at both top and bottom

2 wedges

Clamping blocks to suit

Blocks at ends of form cores

Core

Riser

Core

Tread

Figure 2   *Bottom-side-up elevation of tread and riser.*

1'

Plate 17—Two-Platform Stair Similar to Plan G of Plate 12

Figure 1 shows a two platform staircase with 3-½" square-bottom posts at the start, platform corners, and at the balcony. Stringers are 1" thick and mitered to the risers at the open side. 1" wall stringers are the housed-and-wedged type.

Figures 2 through 4 are perspective elevations showing the posts recessed to receive the stringers. Post 4 is notched at the balcony header.

Plates 18 and 19—Two-Platform Stair with Buttress-Type Housed and Wedged Stringers at the Open Side of the Stair.

This is a buttress stair with 5" hollow bases at the starting tread, corners of the platforms, and at the balcony to receive 3-½" square-bottomed posts. This type of stair has been traditional before the twentieth century.

In **Plate 18, Figure 1** shows the plan, **Figure 2**, the 5" hollow bases with the buttress centered, **Figure 3**, the buttress section, and **Figure 4**, the starting base elevation.

In **Plate 19, Figure 5** shows the lower and upper buttress runs at posts 2 and 3, and **Figure 6**, the upper buttress run to the balcony post 4.

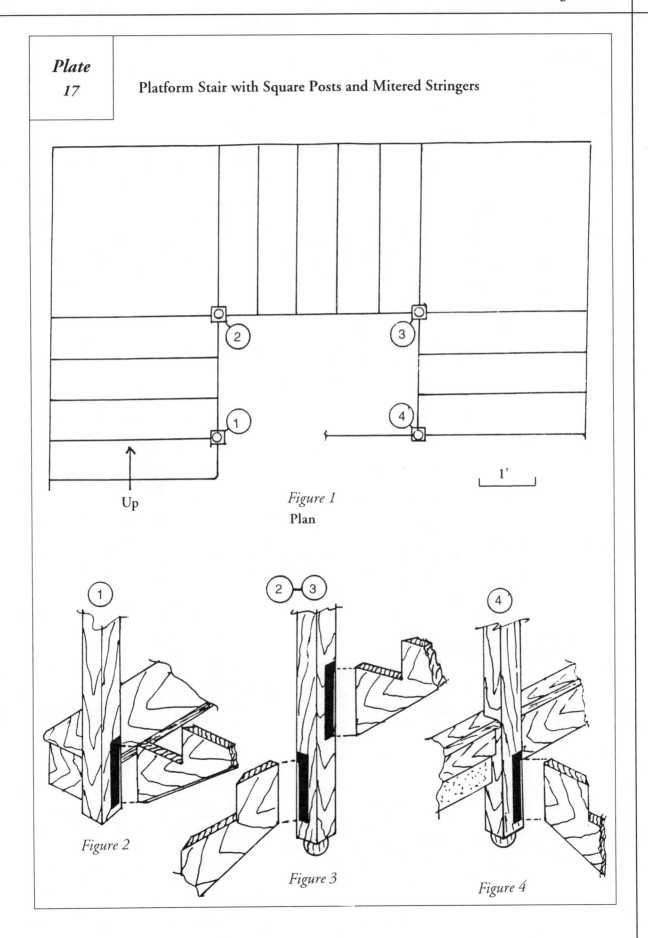

Plate 17

Platform Stair with Square Posts and Mitered Stringers

Figure 1
Plan

Up

1'

Figure 2

Figure 3

Figure 4

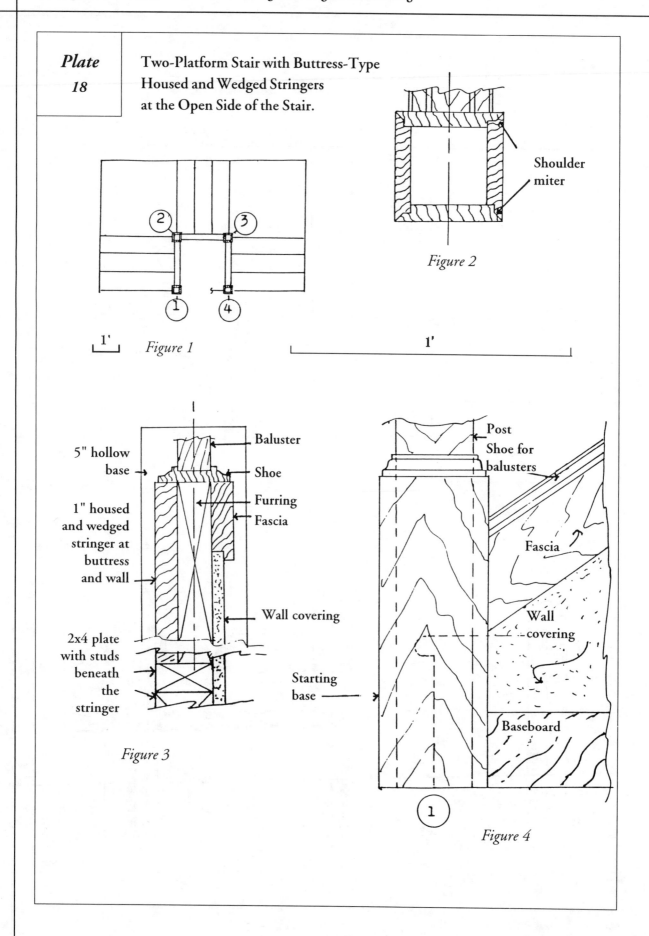

Plate 18 Two-Platform Stair with Buttress-Type Housed and Wedged Stringers at the Open Side of the Stair.

Figure 2

Shoulder miter

1'

Figure 1

1'

5" hollow base

1" housed and wedged stringer at buttress and wall

2x4 plate with studs beneath the stringer

Baluster

Shoe

Furring

Fascia

Wall covering

Figure 3

Post

Shoe for balusters

Fascia

Wall covering

Starting base

Baseboard

Figure 4

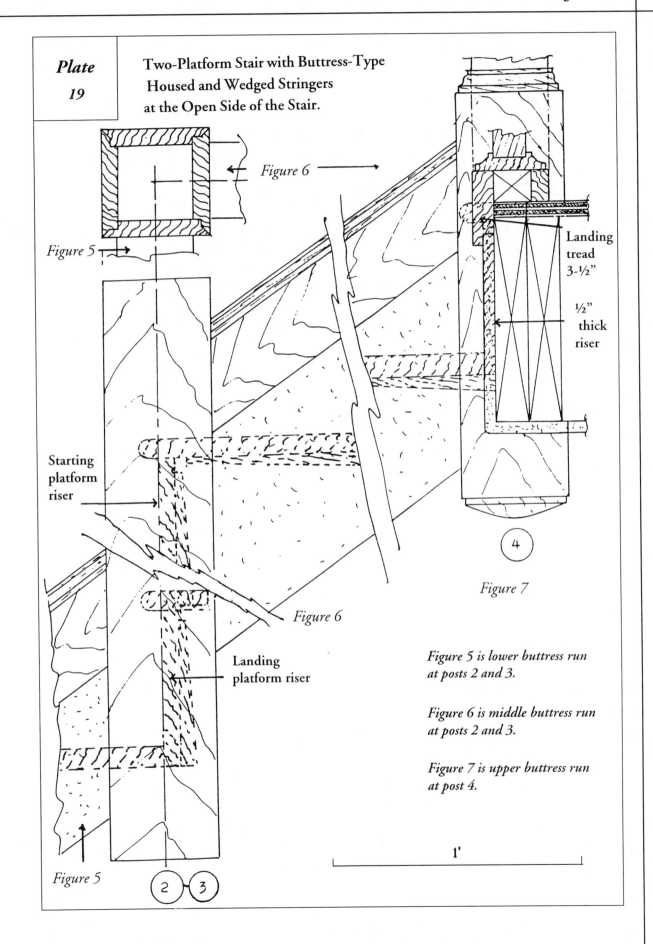

| Plate 19 | Two-Platform Stair with Buttress-Type Housed and Wedged Stringers at the Open Side of the Stair. |

Figure 6

Figure 5

Landing tread 3-½"

½" thick riser

Starting platform riser

Figure 6

Landing platform riser

④

Figure 7

Figure 5 is lower buttress run at posts 2 and 3.

Figure 6 is middle buttress run at posts 2 and 3.

Figure 7 is upper buttress run at post 4.

Figure 5

② ③

1'

Plates 20 through 22—Six-Winder Stair Similar to Plan J of Plate 12, with 1-inch Housed and Wedged Type Stringer Shown in Figure 1, Plate 7

Plate 20 is a plan of the winder section of a six-winder type stair between walls, with runs of square treads. The traveling line from A to B is approximately 15" from the narrow end of the winders. The space between risers at this point approximates that of a normal straight run square tread.

Risers 7 and 8 should be located away from the corner of the "boxed-corner" so that the nosing projections of both the winder and square treads will not encroach into the mitered corner. See **Figure 2** of **Plate 21**.

Figure 1 in **Plate 21** shows the plan of the stringers and boxed-corner.

Figure 2 is the stretch out elevation of the routing of the boxed-corner and connecting stringers.

Plate 22 shows the elevation of the housed and wedged stringer at the outside wall.

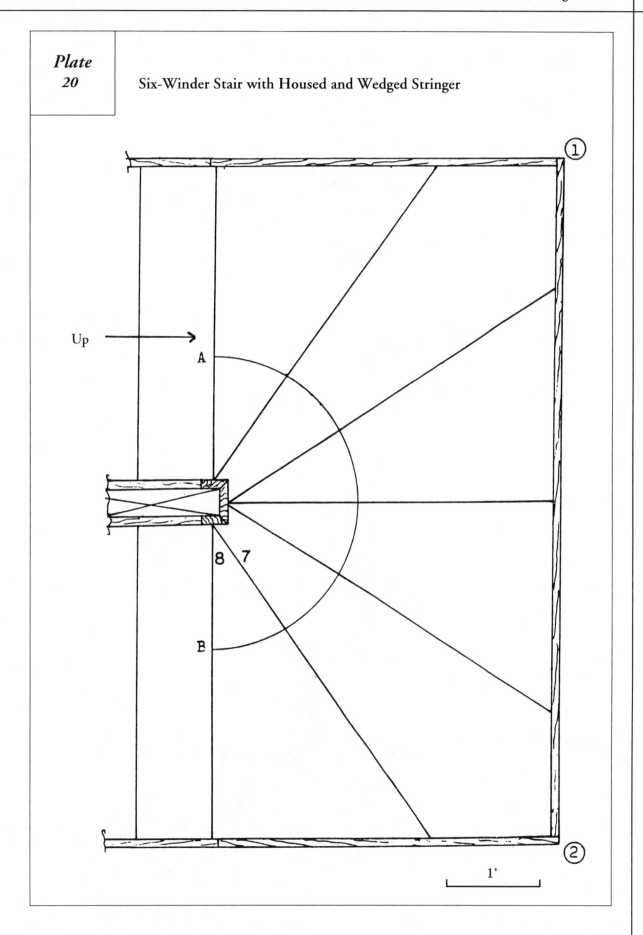

Plate 20   Six-Winder Stair with Housed and Wedged Stringer

Up

A

①

②

8 7

B

1'

Plate 21

Plan and Elevation of 1" Thick Box-Corner Stringer

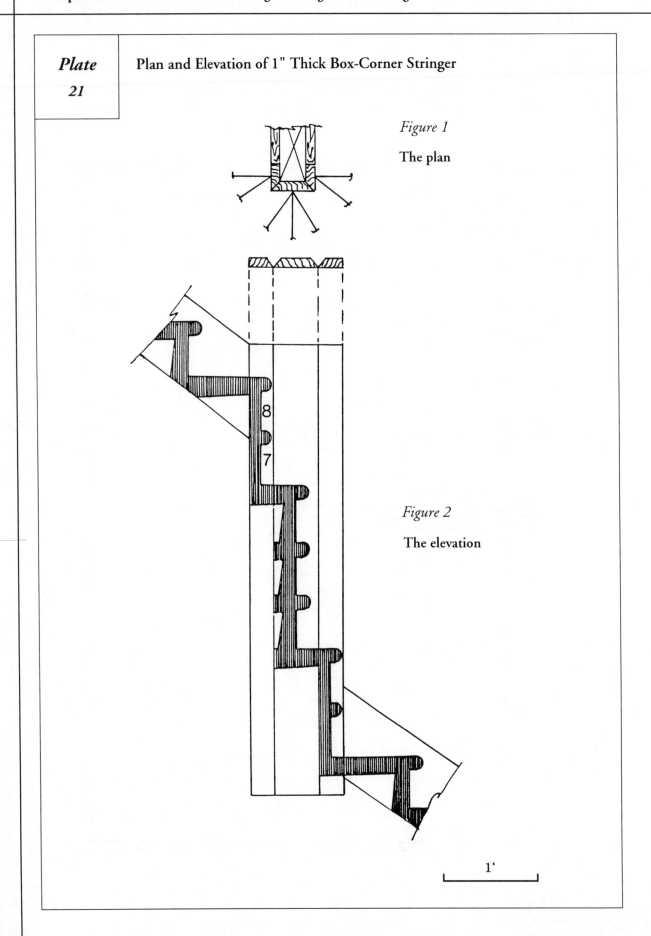

Figure 1

The plan

8

7

Figure 2

The elevation

1'

Plate
22

Elevation of 1" Thick Housed and Wedged Wall Stringers

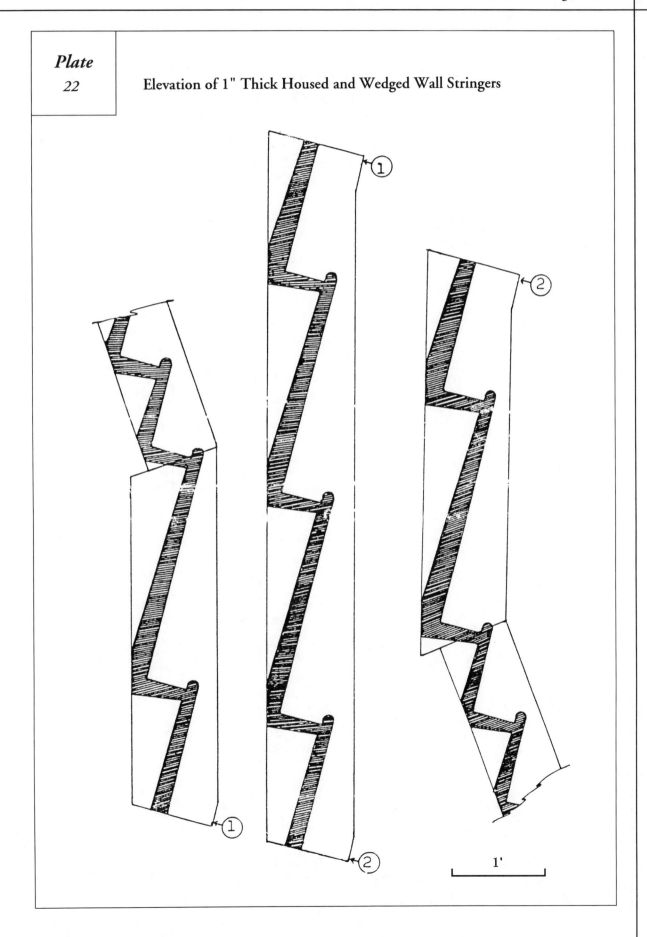

1'

Plate **23**	**Right-Angle Single Platform Stair**
	with 1-Inch Thick Housed and Wedged Stringers

The stair plan shows a wall corner at the stringer near the starting riser. Three possible positions of the corner are shown in **Figures 1, 2,** and **3. Figure 4** is the stringer at the corner of the two straight runs.

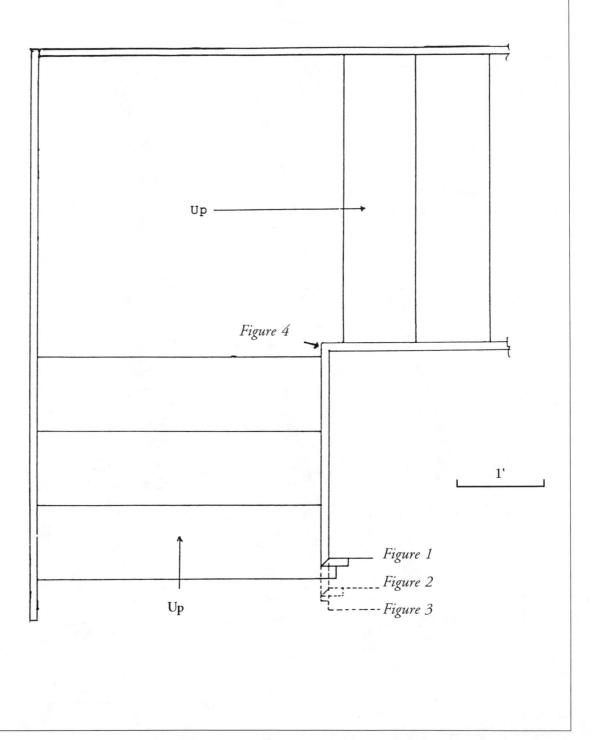

Up →

Figure 4

1'

Up ↑

Figure 1
Figure 2
Figure 3

Up

Plate 24

Right-Angle Single Platform Stair
with 1-Inch Thick Housed and Wedged Stringers

Plate 24 shows the three plans and stringer build-up elevations to properly terminate the stringer in the starting riser locations shown in **Figures 1, 2,** and **3** of **Plate 23.**

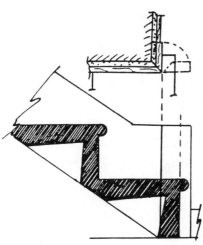

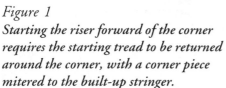

Figure 2
Starting the riser just short of the corner allows the nosing of the starting tread to remain on the straight stringer. Stringer requires a return piece around the corner as in Figure 1, to receive the baseboard.

Figure 1
Starting the riser forward of the corner requires the starting tread to be returned around the corner, with a corner piece mitered to the built-up stringer.

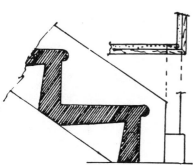

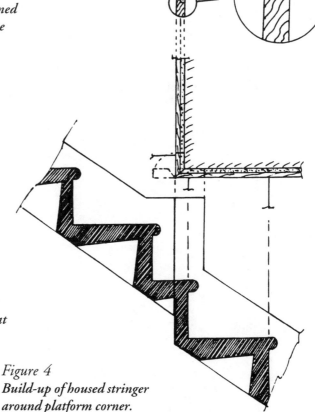

Figure 3
Starting the riser considerably short of the corner allows the stringer to be cut on the plumb, forward of the tread nosing, and still receive the baseboard at the straight stringer.

Figure 4
Build-up of housed stringer around platform corner.

Plate 25—Stair Plan R of Plate 12

This is a winder type stairway with a 3-½" square center post, housed and wedged wall stringers, a straight run mitered stringer at the top of the stairs, an exposed soffit, and the top riser forward of the normal header line so as to allow head room over the starting riser.

In planning this type of stair, the prime consideration is to obtain headroom at the start of the stair. A scale drawing must be made so that the risers can be properly placed in order to be certain of the headroom.

Figure 1 is the proposed stair plan consisting of 10 winder treads and 3 square treads to reach the height of 9'-0", which will take 14 risers, each with a height of about 7-¾". The traveling line around the center post will equal that of a square tread.

Figure 2 elevation shows that in order to obtain a minimum of 6'8" headroom from the floor to under the upper mitered stringer soffit at 18" from the post, the top riser must be moved forward of the normal header line as shown in both figures.

The full-round wall handrail is continuous at the wide side of the stairs. A molded type rail is at the mitered-face stringer side and at the balcony levels, receiving balusters as shown.

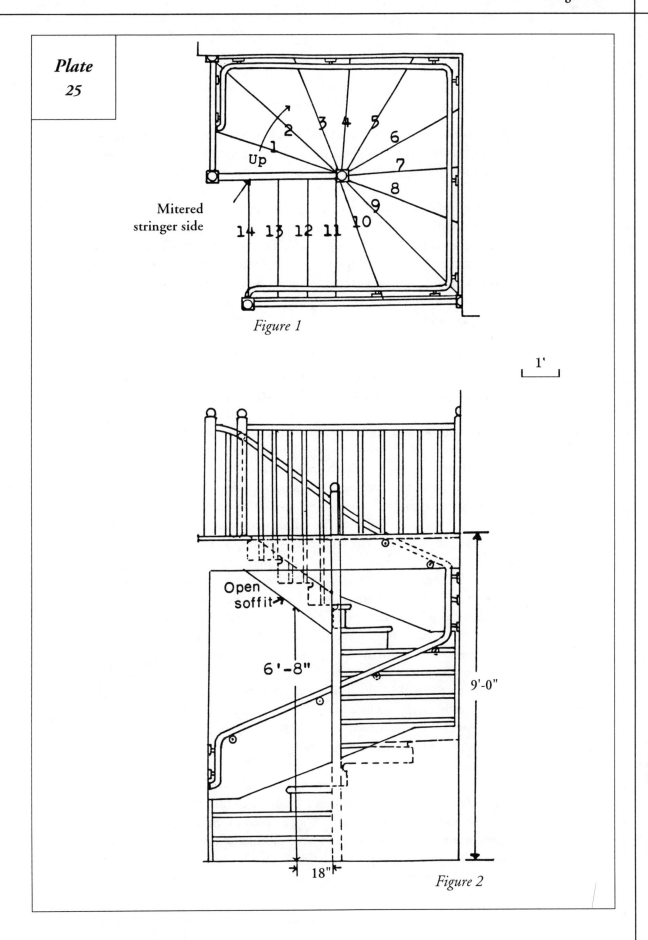

Plate 25

Up

Mitered stringer side

Figure 1

1'

Open soffit

6'-8"

9'-0"

18"

Figure 2

Plate 26—Locating the Starting and Landing Riser from the Starting and Landing Headers of a Straight-Run Stair whenever the Handrail Is To Be Continuous from a Lower Balcony Level to an Upper Balcony Level

Figure 1 of this plate is a plan of a straight run type of stair open at the left side going up. It is to receive a continuous handrailing from the lower balcony level to the upper balcony level. This will require a level-to-rake quarter-turn rail section at the bottom, and rake-to-level quarter-turn rail section at the top, with no other easements involved. For this to occur, the starting and landing risers must be properly placed a specific distance from the starting and landing headers.

At the beginning of this book, I stated the importance for the stairbuilder to have a minimal knowledge of tangent handrailing. **Plate 26,** as in **Plate 13,** is another example of practicable application of that knowledge.

Figure 2 is an elevation of a typical tread and two risers of the plan in **Figure 1.** Let the lower riser be the starting riser and the upper riser be the landing riser. Draw the bottom line of the rail touching the riser tops, then draw the rail centerline and top. Establish the rake rail height as 2'-8". With the distance <u>ab</u> equal to 2'-8", the desired level rail heights of 3'-0 are set up at <u>c</u> and <u>d</u>. The intersection of the rake rail centerline at the level rail centerline locates the vertex lines of the tangents of both quarter-turn rail sections. See **Plates 3** and **13.** The detailed elevation at the top riser shows the squared rail, its centerline, and the width of a typical baluster at the balcony level. The face of the baluster is shown to be in line with the face of a 1" thick fascia applied to the joist, or header. With the top header established, the dimension between the header and the top riser is determined as <u>e</u>. The margin between the header and the centerline of the rail is <u>f</u>. This same margin <u>f</u> is drawn forward of the lower quarter-turn centerline to establish the line of the lower header. There will also be a 1" thick fascia applied to this header, the face of which is in line with the face of the baluster. Margin <u>e</u>, from the lower header to the starting riser, is shown to be a different dimension than margin <u>e</u> from the landing header to the top riser face.

Figure 3 shows the plan of the rail. The starting and landing riser positions are transferred from the elevation of **Figure 2** to this plan, showing their dimensions from the header as <u>e</u>. While a plan of this nature bears out the importance for the stairbuilder to have minimal knowledge of the tangent system of handrailing, it is equally important for the architect/designer, for it is they who draw the plans and allot the stairwell dimensions.

Plate 26

Starting and Landing Risers Located by Rail Layout

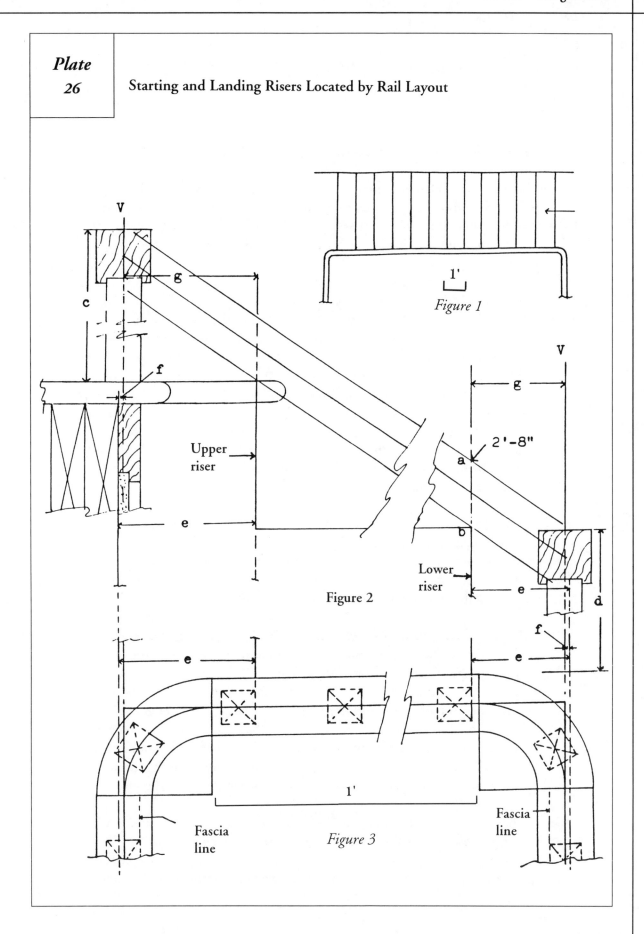

Figure 1

Upper riser

Lower riser

Figure 2

Fascia line

Fascia line

Figure 3

2'-8"

Plate 27—Locating the Riser Positions of a Mitered Stringer Platform-Type Stair with Equally Spaced Balusters

Figure 1 is the plan of the stair showing two straight runs of treads and risers to the platform. <u>a</u> is the distance of 10" between risers. The landing and starting risers at the platform are to be positioned so that there will be a 5" center-to-center spacing between the balusters throughout.

In **Figure 2,** the two runs of mitered stringers will meet at right angles instead of curving with the handrail. A 4" centerline radius for a rail width of up to 2-½" will not create an obtrusive S curve effect at the concave side of the rail, and will not cause the nosing line of the curved landing tread to unreasonably project from the right-angle corner of the intersecting stringers. Since the distance between square tread risers is 10", a baluster spacing at two per tread is to be 5" on centers. Establish a baluster at the center of the quarter turn, and set 5" centered balusters at each side. This will then locate the landing and starting risers as <u>b</u> and <u>c</u> the distance from the vertex to the risers. However, should the stairwell dimensions dictate riser positions other than shown, the handrail can still be satisfactorily made even though the easing of the rail through the turn will not be as graceful.

If the stringers in this plan were the buttress type, then the balusters could be spaced independent of the risers. The risers could then be located as shown in **Plate 13,** where the distance between risers along the plan tangents and the rail centerline equals one square tread. The pitched tangents would then be in line with the normal centerline pitch of the straight rail.

Because of the equal baluster spacing in this stair, there must be a slight adjustment in the turn section of the rail in order to align with the lower and upper straight rail beyond the turn. In **Figure 3,** the stretch out elevation of the plan tangents and risers shows that the pitch of the normal straight rail centerlines of both lower and upper straight runs do not meet at the vertex line. Both tangents are then drawn through the center of the difference as shown. **Plate 102** shows how to compensate for such a condition in making a climbing-turn handrail section. The distance <u>d</u> is adjusted at the surface of the butt joint along the bevel line, while <u>ef</u> is the height through the turn.

The butt joints of the rail sections are located beyond the spring lines whenever possible, so that the rail thickness of the adjoining straight rail will not be within the curve. In this particular section, ample length is allowed beyond the spring lines so as to make a more graceful transition from joint to joint.

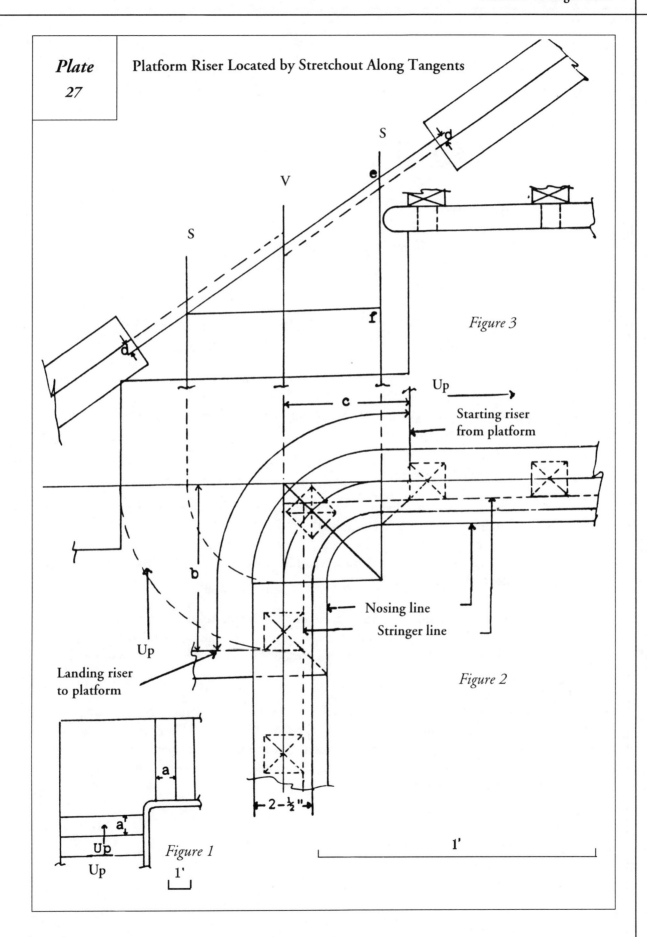

Plate 27

Platform Riser Located by Stretchout Along Tangents

S

d

e

V

S

f

Figure 3

c

Up

Starting riser from platform

b

Nosing line

Stringer line

Up

Figure 2

Landing riser to platform

2–½"

a

a

Up

Figure 1

Up

1'

1'

1'

Section II

Curved Stairs

This section is concerned with the following:

1. Curved stringers.

2. The basic principles of tangent handrailing for making incline-turn handrail sections.

3. Pitch and level tangent combinations in handrailing.

4. Equally spacing balusters throughout the curved stair stringers in order to locate riser positions.

5. Making curved stringers.

Making incline-turn handrail sections and solving handrail problems will be thoroughly dealt with in **Section III**.

Ten Examples of Stairs with Curved Line Construction: The Basic Principles of Tangent Handrailing

Tangent handrailing is a precise method of making solid wood handrail sections that both incline and turn, as opposed to making such handrails by laminate-gluing thin members together, or by trial-and-error fitting of blocks to suit.

The next few **Plates 29** through **32** show preliminary principles in making incline-turn handrail sections with the tangent method. They are the basis on which this tangent system is founded. Presenting them at this time, rather than in **Section III**, will enable the stairbuilder to better plan and design curved stairs involving such handrailings.

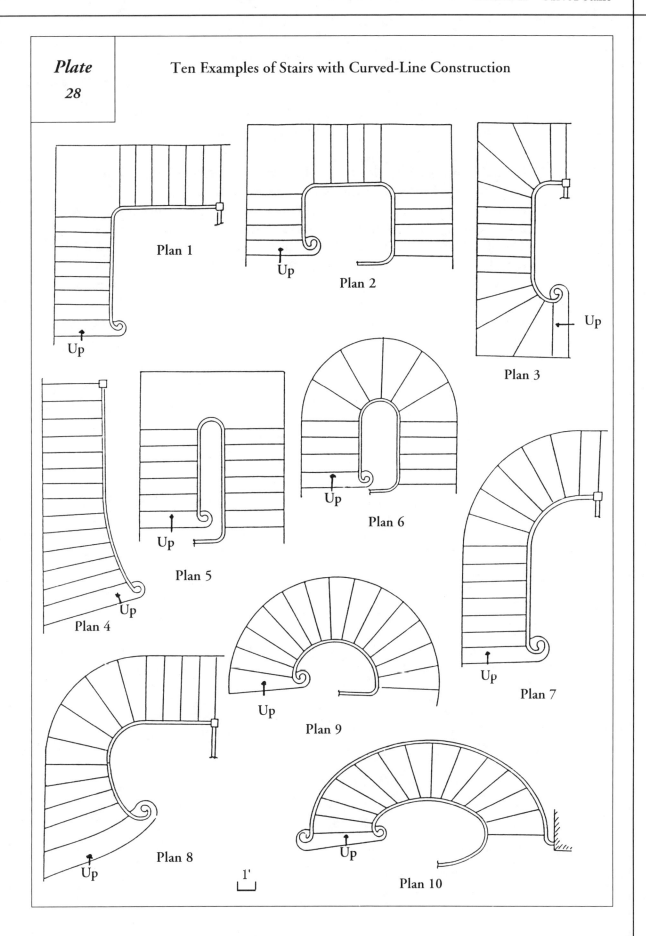

Plate 28

Ten Examples of Stairs with Curved-Line Construction

Plan 1

Up

Plan 2

Up

Plan 3

Up

Plan 4

Up

Plan 5

Up

Plan 6

Up

Plan 7

Up

Plan 8

Up

Plan 9

Up

Plan 10

Up

1'

Plate 29—Simple Climbing Turn Handrail Section

Plate 29 shows an example of a simple twisted handrail section inclining from level to rake. **Figure 1** shows a level quarterturn section with tangents AB and BC. **Figure 2** is a side view of **Figure 1** showing rail thickness. **Figure 3** is an elevation of tangent BC of **Figure 1,** while tangent AB remains level. The rectangular shape of the rail at A determines the block thickness required to make this rail. The bevel at D is applied through A as a plumb line. **Figure 4** is an isometric view of the rail of **Figure 3** after the sides, top, and bottom surfaces have been dressed to size. The following pages in **Section III** will show the step-by-step procedure for making this and other possible climbing-turn handrail sections.

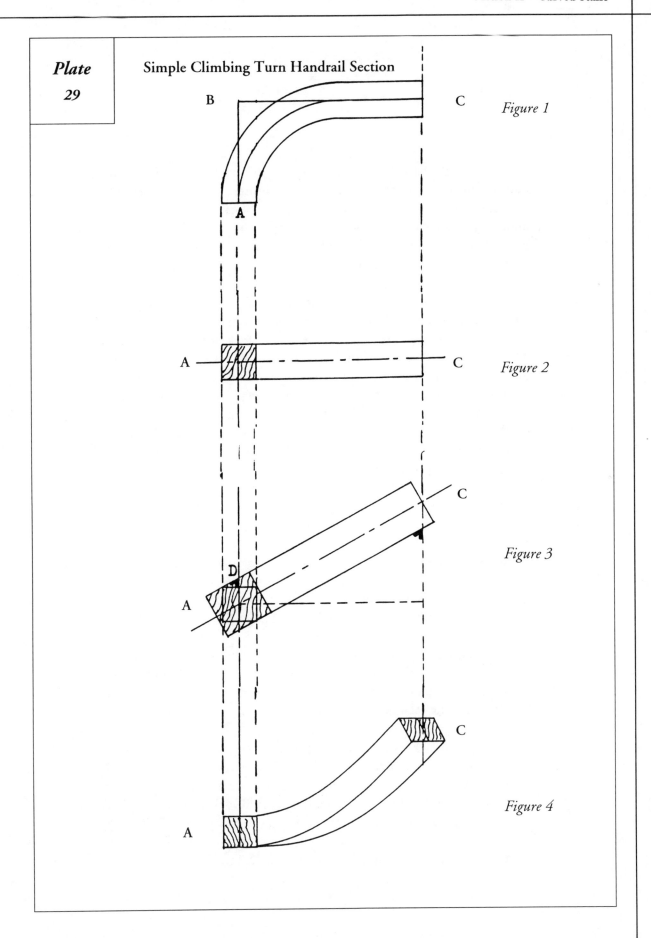

Plate 29

Simple Climbing Turn Handrail Section

B   C   *Figure 1*

A

A   C   *Figure 2*

C

D

A   *Figure 3*

C

A   *Figure 4*

Plate 30—The Prismatic Solid

Solving all of the basic handrail problems in this book is the result of the study of simple blocks of wood with right, obtuse, and acute angled sides simulating tangents of a handrail section. The blocks are cut in length to simulate the tangents at an incline, as in an incline-turn handrail section. The right-angle block shown in **Plate 30** is an example of a quarter-turn handrail section with equally pitched tangents.

Figure 1 shows a wood block example of a square section cut on a diagonal from C' to A, the face of the oblique cut being a parallelogram AB'C'O'. The dotted quarter circle shown at the base of the block, from A to C, simulates the plan of a handrail centerline, with AB and BC respective plan tangents to the curve at A and C, which are termed spring-line points. B is the vertex, or meeting point of the two tangents. The oblique plane shows the simulated pitch of the plan tangents inclining from left to right, the lower tangent pitched as AB', and the upper tangent pitched as B'C'. Since BB' and OO' are equal height, both tangents are equally pitched, therefore, equal length. The total height through the turn from A to C is CC'.

Figure 2—This is the same solid as in **Figure 1** looking from the shaded side AOO', showing the oblique plane as AB'C'O'.

Figure 3 shows the oblique plane of **Figure 1** rotated along the line of tangent B'C' as an axis so that the oblique plane will lie on a flat surface. This means that point A is rotated square to the upper tangent B'C'. Therefore, the angle of the elevated tangents shown on a flat surface is AB'C'.

Figure 4 illustrates all six sides of the prismatic solid of **Figure 1** unfolded on a flat surface to simulate a template. A template such as this can be made of any of the basic handrail problems of **Plates 73** through **87** if it is desired to further prove the accuracy of the layouts.

Figure 5 is a surface layout to find the angle of the elevated tangents employing the principles shown in **Figures 2 and 3**. ABCO is the plan parallelogram. AB and BC are the plan tangents. CC' is the height CC' of **Figure 1**. A"B is drawn perpendicular to upper tangent B'C'. B'C' equals A'B'. A" and B' are connected to form the angle of the elevated tangents A"B'C'.

D, x, and y of **Figures 4 and 5** are fully explained in 'The Bevel,' of **Plate 64**.

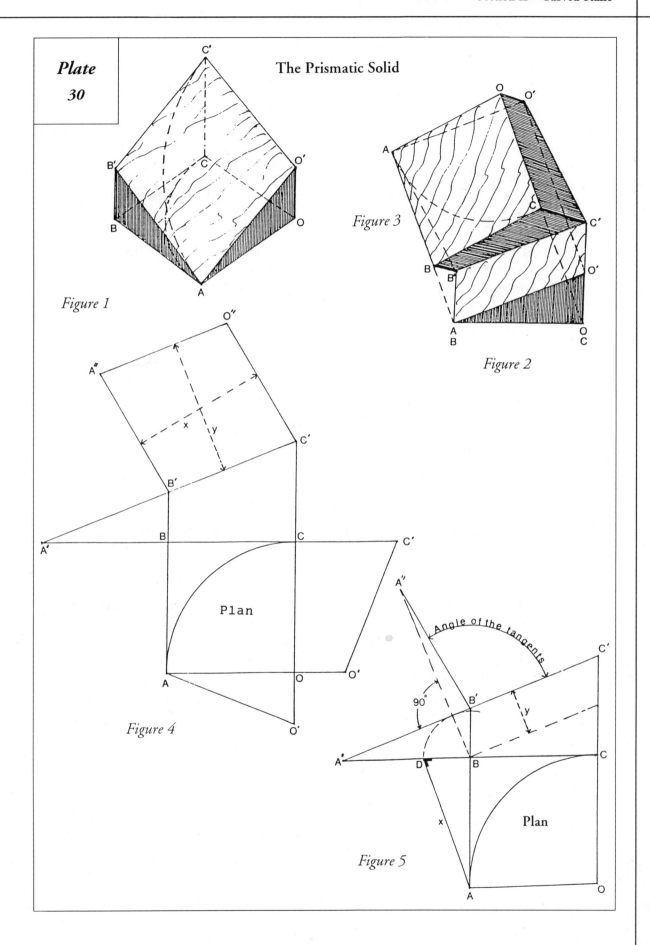

Plate 30

The Prismatic Solid

Figure 1

Figure 2

Figure 3

Figure 4

Plan

Figure 5

Plan

Angle of the tangents

90°

Plate 31—The Cardboard Model

The saying "a picture is worth a thousand words" is graphically illustrated by a cardboard model made up from a template such as this to demonstrate the principles of tangent handrailing.

The cardboard model is the counterpart to the prismatic solid as the foundation for the system of tangent handrailing as presented in this treatise. All of the handrail problems shown in the chart in **Plate 72** have been proven by a cardboard model such as this.

Plate 31 is a similar template of the prismatic solid of **Figure 4, Plate 30**. The beginning handrail maker should make a cardboard model of this template. By folding the model so that the elevated tangents A"B' and B'C' lie directly over AB and BC of the plan, the oblique parallelogram lies over its plan parallelogram ABCO. With the supporting sides folded as shown, the model simulates the prismatic solid of **Figure 1, Plate 30**.

Section III explains the importance of the ordinate line BO and the minor axis line B'O' for finding the required curved pattern of the plan rail width as it is projected along the surface of the oblique plane parallelogram A"B' C'O'. The pattern to be found at the angle of the elevated tangent's angle A"B'C' is referred to as the face mold, which will be further explained in subsequent plates.

Although this model is for a quarter-turn section showing equally pitched tangents, where heights BB' and OO' are equal, unequally pitched tangents will show that these heights will not be equal. This is shown in **Plates 48** through **52**. Consequently, the ordinate BO, and the minor axis B'O', both indicated by a small circle through the lines, will not strike exactly at respective vertices B and B'. The importance of the directions of the ordinate and minor axis with respect to vertices B and B' is explained in **Plates 58** through **62**.

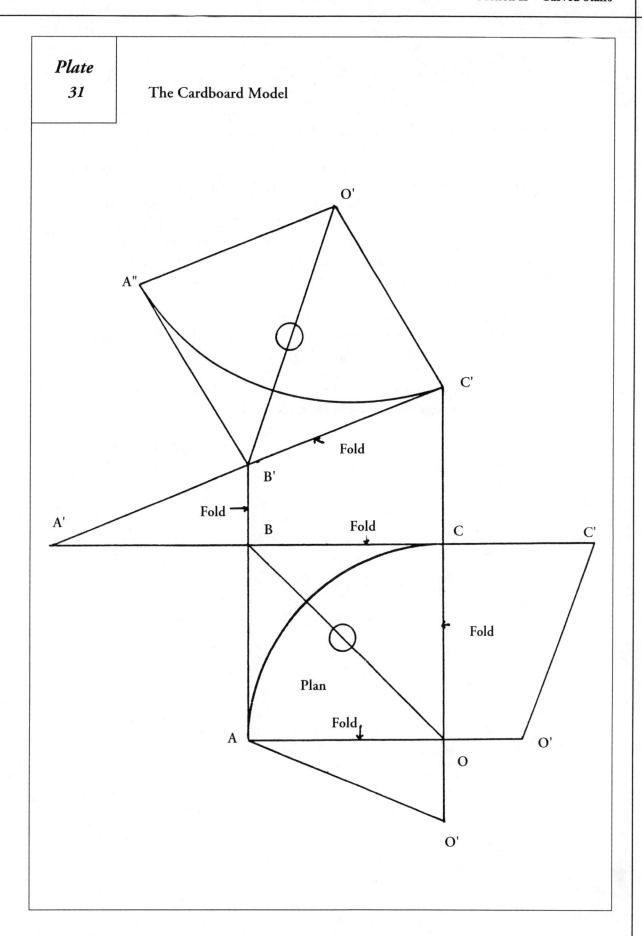

Plate 31 The Cardboard Model

Plate 32—Pitch and Level Tangent Combinations

This plate shows quarter-circle plans with the plan tangents stretched out and extended into either elevation or with one tangent remaining level. These combinations of tangent pitches and level conditions are also applicable to obtuse and acute angle plans. Remember, V equals the vertex line, the meeting point between the two tangents forming the angle of the tangents; S is the vertical extension of the spring line, the point where the tangent touches the curve; and t refers to either the level or pitched tangents.

Figure 1 shows the quarter-turn with equally pitched tangents; **Figure 2**, a short lower pitched tangent; **Figure 3**, a short upper pitched tangent; **Figure 4**, a lower level tangent; and **Figure 5**, an upper level tangent.

The intersection of the risers at the plan tangents will determine the pitch of the elevated tangents as shown in **Plate 13**.

In the planning of a curved staircase, baluster spacing is a prime consideration. **Plate 33** will further explain the importance of baluster spacing to riser locations.

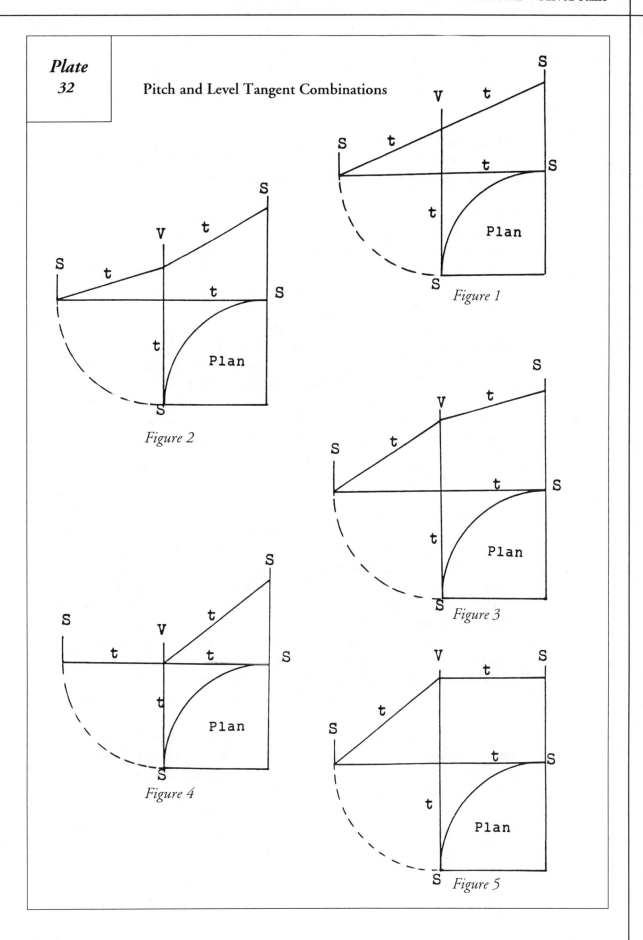

Plate 32

Pitch and Level Tangent Combinations

Figure 1

Figure 2

Figure 3

Figure 4

Figure 5

Plate 33—Baluster Spacing in Four Quarter-Circle Stairs Involving Two, Three, and Four Winders Plus Straight (Square) Treads. The Stringer Is a Mitered Type at the Handrail Side of the Stair.

In designing curved stairs, the narrow ends of the winding treads must comply with local or state building codes. The spacing between risers at this narrow end, in most instances, should not be less than 6". Assuming the minimum of 6", the stairbuilder/designer must also consider from an aesthetic viewpoint that all balusters supporting the handrail should be equally spaced along the treads, or so nearly equally spaced that there would be no noticeable difference. If there are also square treads in the curved stair, the spacing between risers of a winder tread, along the face of the mitered stringer, is determined by the spacing of balusters at a square tread. If the square tread is 12" wide between risers, receiving two balusters, then the winder treads should be spaced 6" between risers, and will receive one baluster. If the square treads are 10" wide, they must receive three balusters, making the center-to-center spacing between three balusters 6-⅝", which will also be the riser-to-riser spacing at the winder treads receiving two balusters. In this manner all balusters will be equally spaced.

Figure 1—In this quarter-circle plan, two balusters at 5" centers are desired at the 10" square treads. Two winder treads are required at the quarter-turn with a minimum spacing of 6" between risers at the face of the mitered stringer. With risers placed at the spring lines, and the face of a baluster in line with the riser, the spacing of 5" center-to-center balusters requires a radius of 9-½" to the rail center. The distance between the spring-line risers and the winder tread riser is 7-7/16". Assuming the nosing projection to be 1" forward of the winder riser, as at x, the 5" center-to-center spacing between balusters on the lower winder tread would leave approximately 1-½" from a baluster center to the nosing line y, enough for say a 1-½" square baluster to clear the nosing by ¾". All baluster spacings will then be equal.

Figure 2—This three-winder stair plan will require three balusters per 10" square tread, or 6-⅝" between two baluster centers. The centerline radius is calculated to be 13" in order for the winder treads to be a minimum of 6-⅝" from riser-to riser so as to include the two equally spaced balusters to a winder tread. The walking line, approximately 12" to 15" from the handrail, should be neither less than a normal square tread width nor greater than one-third the width of a square tread.

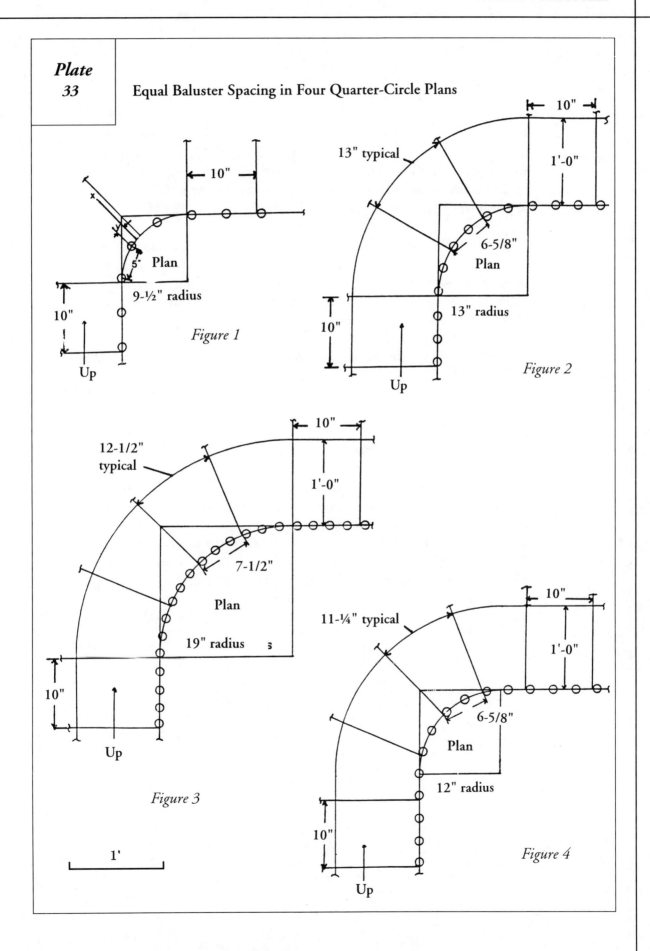

Plate 33

Equal Baluster Spacing in Four Quarter-Circle Plans

10"

13" typical

1'-0"

6-5/8"
Plan

5'
Plan

9-½" radius

10"

13" radius

10"

Figure 1

Up

10"

Figure 2

Up

12-1/2"
typical

10"

1'-0"

7-1/2"

Plan

19" radius

10"

11-¼" typical

10"

1'-0"

6-5/8"

Plan

12" radius

10"

Up

Figure 3

1'

Up

Figure 4

Figure 3—This is a plan of a four-winder quarter-circle stair showing four balusters at a square tread with 7-½" centered between four balusters. The four winder treads are then spaced 7-½" between risers along the centerline of the balusters, requiring a 19" radius. The traveling, or walking, line is approximately 12-½" from the centerline of the rail.

Figure 4—This is also a plan of a four-winder quarter-circle stair with winder risers beyond the spring lines of a predetermined 12" centerline radius. Since there are 10" square treads with three balusters, the spacing between the three balusters is 6-⅝". Therefore, the spacing between risers at the winders is also 6-⅝". The walking line is approximately 11-¼".

The tangent pitches for the handrails for these four plans are shown in **Plates 120** and **121**.

Plates 34 through 39—Other Stair Plans Showing Equal Baluster Spacing.

Plate 34 is a segment of a five-winder stair showing equal baluster spacing. The risers of the winder treads in this stair must be a minimum of 6" apart along the face of the mitered stringer. In order to carry out the same 5" center-to-center baluster spacing of the 10" square treads, the face of all balusters placed on the winder treads will not necessarily be in the desired line of the face of the winder risers since equal spacing of the balusters is mandatory.

Plate 34

Baluster Face Not Necessarily In Line with Risers

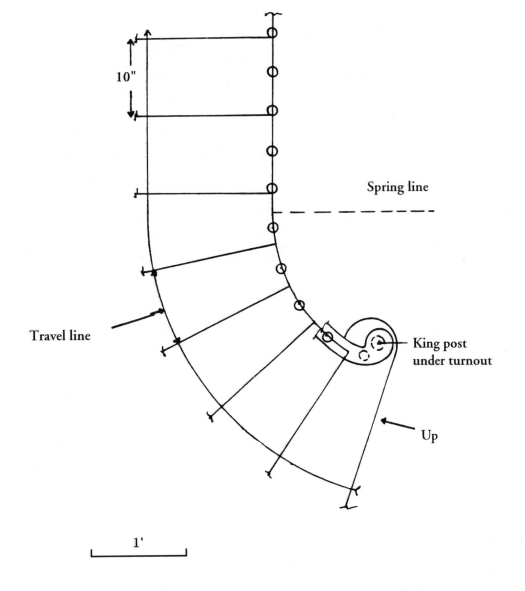

10"

Spring line

Travel line

King post under turnout

Up

1'

Plate 35 shows **Plan 3** of **Plate 28** with the baluster spacing to be either 5" or 6" on centers. If the building code allows 5" riser-to-riser winder spacing in this stair, the traveling line and 10" square tread widths would be the same with all balusters spaced 5" on centers. The centerline radius of the quarter-turns would be 16". Likewise, if the code requires winder risers to be a minimum of 6" apart along the face of the mitered stringer, the traveling line and the square treads would be approximately 12". The quarter-circle radius to the centerline would then be 20".

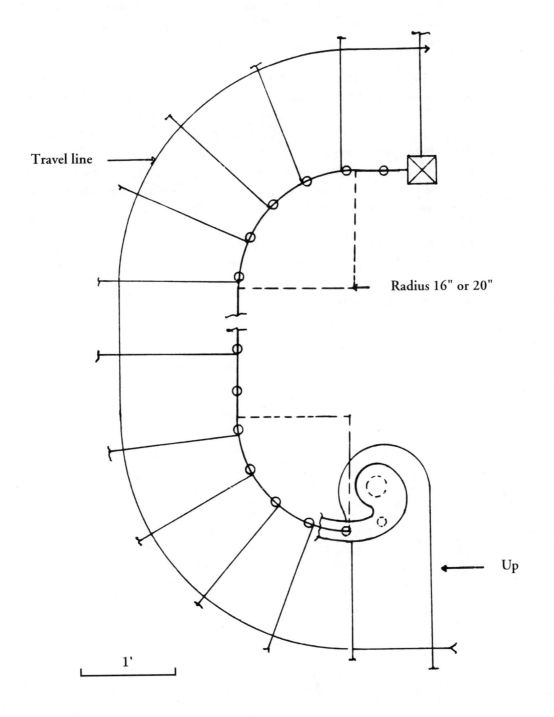

Plate 35

Plan 3 of Plate 28 Showing Baluster Spacing

Travel line

Radius 16" or 20"

Up

1'

Plate 36 shows baluster spacing of **Plans 4, 5,** and **5A** of **Plate 28.**

Plan 4 shows the spacing between risers at the curve of the stair to be equal to that of a square tread. Therefore, all balusters are shown equally spaced at two per tread.

Plan 5 is a stair with a half-circle turn at a platform showing four balusters at the platform. The square treads are to be 10". Therefore, the baluster spacing is to be 5" on centers. Baluster spacing in this stair will dictate the riser locations in the stair plan. Four balusters equally spaced at 5" centers between the spring lines of the half-circle, indicates the radius to be 6-⅜" to the baluster centerline. With a baluster centered at the half-circle, mark two 5" spaces at each side to locate landing and starting riser locations, showing the face of the baluster to be in line with the face of the riser.

Plan 5A is a similar platform type stair as **Plan 5,** with the radius and baluster placements the same. However, there are only two balusters at the platform. The landing and starting platform risers are located at the balusters at each side of the center baluster.

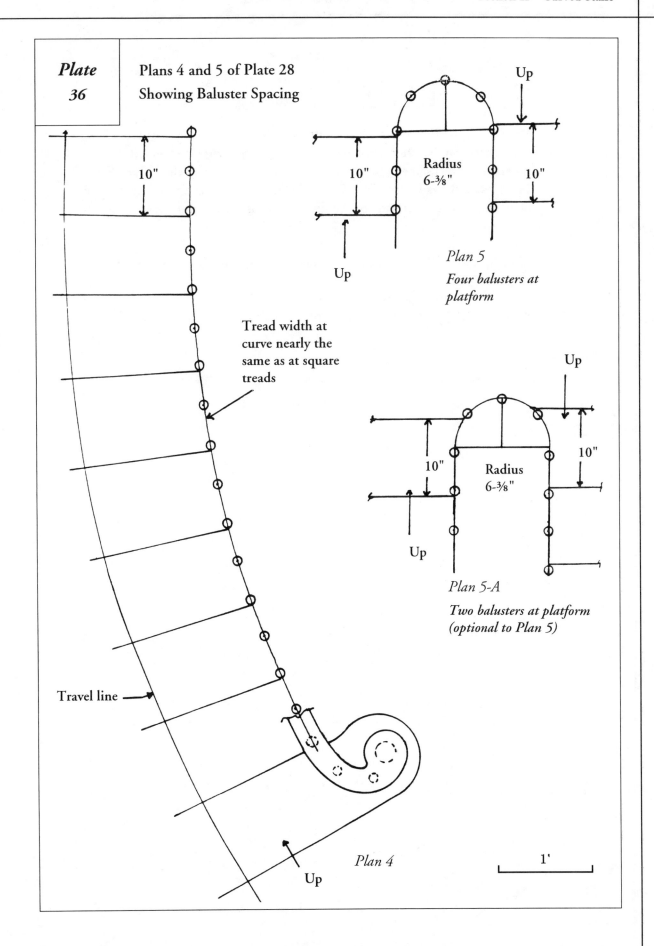

Plate 36

Plans 4 and 5 of Plate 28
Showing Baluster Spacing

10"

Up

10"   Radius
6-⅜"   10"

Up

Plan 5

Four balusters at platform

Tread width at curve nearly the same as at square treads

Up

10"   Radius
6-⅜"   10"

Up

Plan 5-A

Two balusters at platform (optional to Plan 5)

Travel line

Plan 4

Up

1'

Plate 37 shows Plans 6 and 7 of Plate 28. The 11-½" square treads in these plans show two balusters spaced at 5-¾". If the spacing between winder tread risers can be this same 5-¾" then the radius for the half-circle in Plan 6 is 11-½", and the radius for Plan 7 is 22". The winder riser spacing in these two plans can be placed outside of the spring lines as shown and the square tread increased accordingly.

Plate 38 shows Plan 8 of Plate 28 with three balusters at each square tread, and two balusters at the winder treads.

Plate 39 is that of Plan 9 of Plate 28 showing a single baluster at each winder tread.

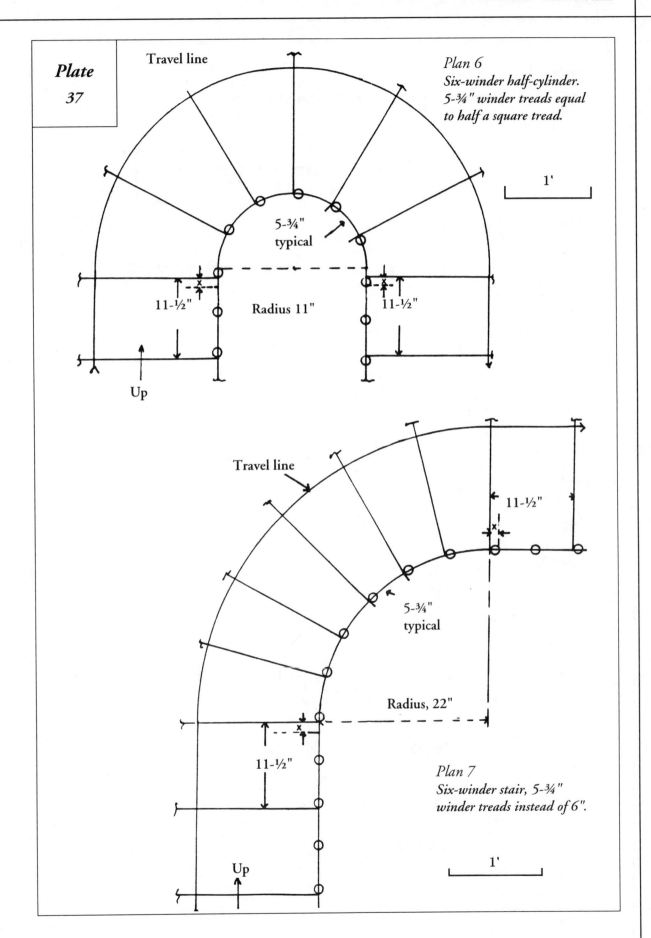

Plate 37

Travel line

Plan 6
Six-winder half-cylinder.
5-¾" winder treads equal
to half a square tread.

1'

5-¾"
typical

11-½"   Radius 11"   11-½"

Up

Travel line

11-½"

5-¾"
typical

Radius, 22"

Plan 7
Six-winder stair, 5-¾"
winder treads instead of 6".

11-½"

Up

1'

Plate 38

Plan 8 of Plate 28 Showing Baluster Spacing

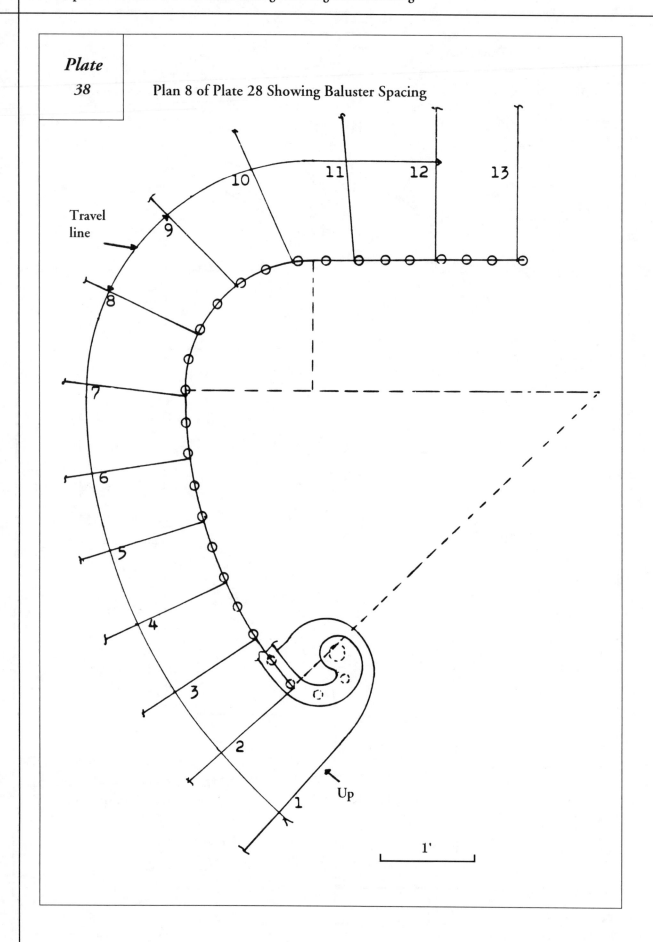

Travel line

Up

1'

Plate 39

Plan 9 of Plate 28 Showing an Elliptical Stair Plan

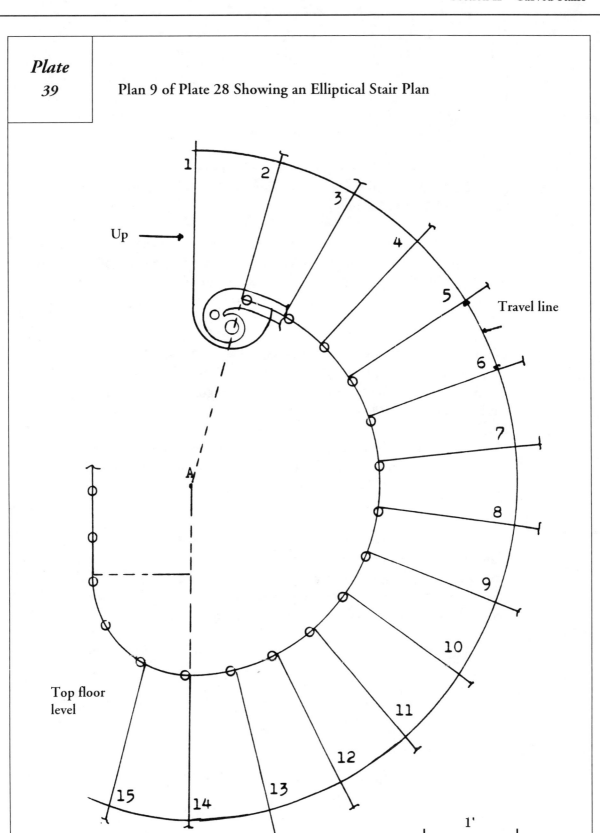

Up →

1 2 3 4 5 6 7 8 9 10 11 12 13 14 15

Travel line

A

Top floor level

1'

Plate 40—Plan 10 of Plate 28—The Elliptical Staircase

While the method of drawing an ellipse with a trammel has been shown in **Plate 2,** for all practical purposes, the elliptical stair shown in this plan is constructed from radius points.

First, in order to find radius points to closely approximate the elliptical curves for this plan, one-fourth of the smaller ellipse for the inside curve of the stair is actually drawn with the trammel. Once drawn, radius points to approximate the curve can be located along both the major and minor axis and at the radius lines.

To begin, let AB be the line of the major axis, with XX the major diameter of the greater ellipse. With the stair width to be CD from C throughout the stair, the minor axis length OD and the major axis length OE are established. The curve is drawn as in **Plate 2, Figure 3** with the trammel.

The elliptical curve of the quarter section can now be closely approximated by trial-and-error radius point locations, as at F on the minor axis line, G at the radius line, and H on the major axis line. The right half of the plan can now be drawn from these radius points.

By transferring the radius point to the left side, the opposite half of the stair can likewise be drawn. With risers 2 and 15 established along the line of the major axis, the remaining risers, 3 to 14, can then be drawn equally spaced at the mitered-face type stringers. Single balusters are shown at the narrow tread ends, and three balusters per tread at the wide ends.

Plan tangent layouts and the pitch of the tangents over the risers will be shown for this and preceding curved stair plans in **Section III.**

Plate
40

Plan 10 of Plate 28 Showing an Elliptical Stair Layout

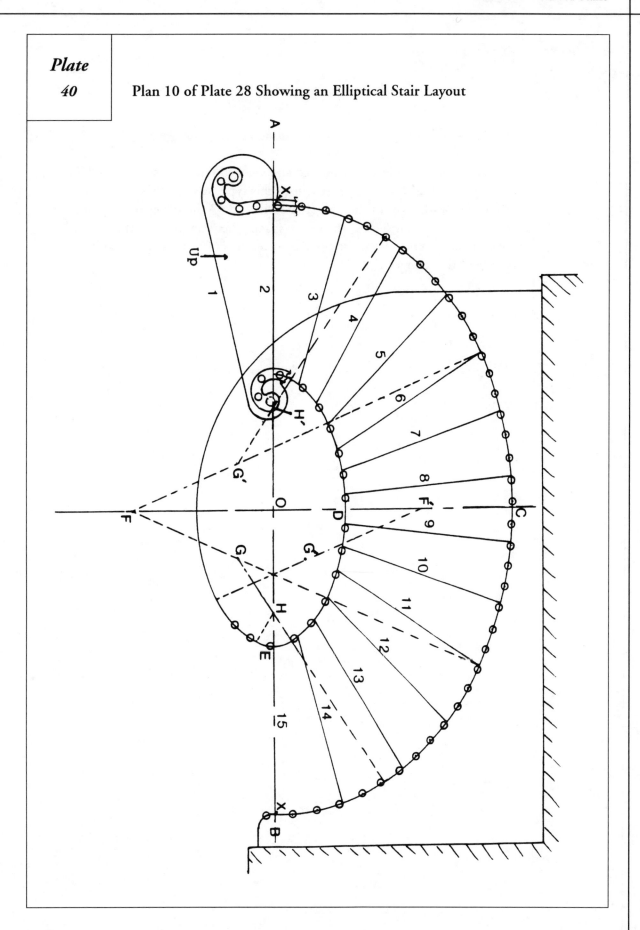

Plate 41—15 Riser Stair with Single Radius Throughout

In this circle stair, there is a 1" housed and wedged type wall stringer and an open-end mitered-face stringer. A stair of this type generally has an exposed soffit underneath most of the stair (**Plates 42** and **43** show construction for an exposed soffit stair such as this). The spacing between risers at the mitered-face stringer is to be 6", showing one baluster per tread, making the radius to the face of the stringer 17-3/16".

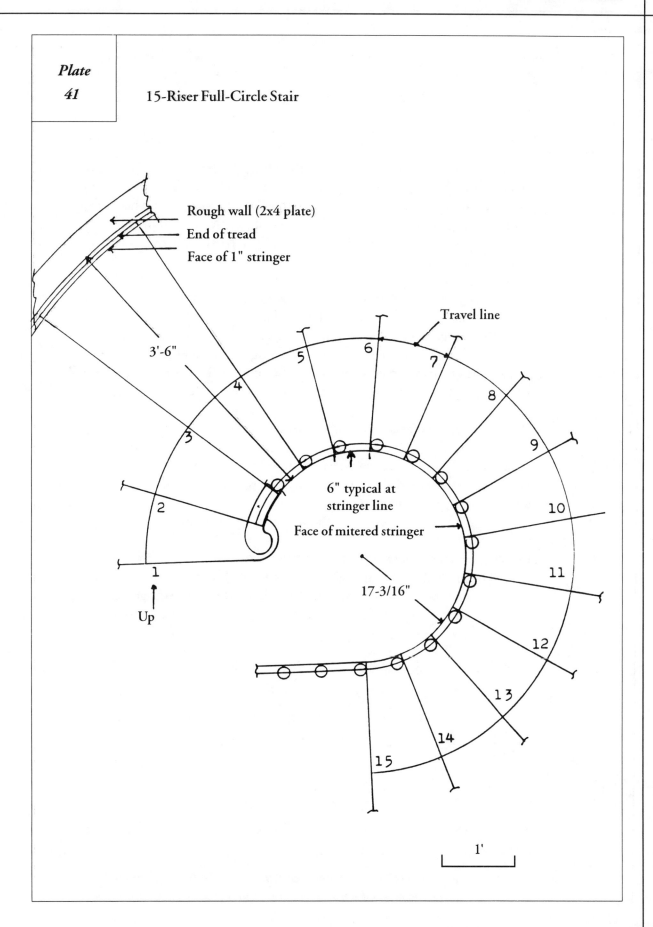

Plate
41

15-Riser Full-Circle Stair

Rough wall (2x4 plate)
End of tread
Face of 1" stringer

Travel line

3'-6"

6
5
7
4
8
3
9
2
10
1
11
12
13
14
15

6" typical at
stringer line

Face of mitered stringer

17-3/16"

Up

1'

Plates 42 and 43—Deep Risers for Self-Supporting Circular Stairs

Certainly all staircases, whether they are made up of straight runs or involve curved stringers, must be soundly constructed to withstand great weight; especially curved, or circular staircases with a warped open soffit where one, or both, stringers are freestanding. The general practice of reinforcing such stairs is to use a combination of circular carriages, cross timbering, and thickened stringers.

Freestanding stair stringers, whether they are mitered or the buttress type, are likened to the side members of a ladder. Whenever the ladder lies at a low pitch, its side members will sag much more than if the ladder were to lie at a steeper pitch. In this same respect to stair stringers, the tendency of the lower-pitched stringer of a curved staircase is to flex greater than the steeper-pitched stringer.

During my many years of making circular staircases, I have used a tried-and-true method of negating the sagging tendency of freestanding curved stringers without the use of central carriages, timbering, and excessive stringer thickness. The method also provides a natural warped soffit beneath the stair to which wire mesh and plaster can be applied. I have termed the method the "deep riser" method. It resembles the web of a steel I-beam.

In a circular stair, where both stringers are freestanding, the sagging tendency of the low pitched outside stringer is negated whenever the deep risers are firmly secured to both stringers, especially when the steeper stringer is a minimum of one and one half times the thickness of the outside stringer.

Deep risers counteract the tendency to sag in this self-supporting circular stair, which also features a turned-and-panel balustrade.

Figure 1 of **Plate 42** is a plan of a 13-riser half-circle stair with a housed and wedged type wall stringer and a mitered-face type open-end stringer with an exposed plaster soffit. This stair is to receive the deep riser. **Figure 2** is an elevation perspective of the stair showing the deep risers from the eighth riser and above. With the risers firmly secured to the wall stringer, the risers act as cantilevered supports for the mitered face stringer.

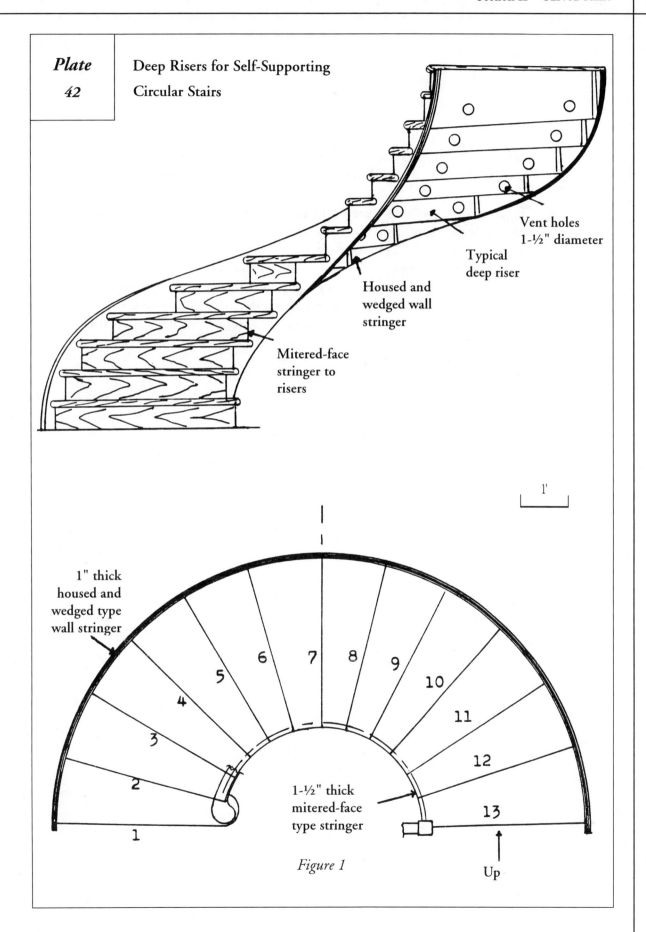

Plate 42

Deep Risers for Self-Supporting Circular Stairs

Vent holes 1-½" diameter

Typical deep riser

Housed and wedged wall stringer

Mitered-face stringer to risers

1'

1" thick housed and wedged type wall stringer

1-½" thick mitered-face type stringer

1 2 3 4 5 6 7 8 9 10 11 12 13

Up

Figure 1

Plate 43 shows perspective details of both the mitered and housed wall stringers in **Figures 3** and **4,** with equal widths shown as A to the stringer bottom. B and C are the deep riser widths from the tread bottom.

Figure 5 is a see-through plan of the risers and stringers with the tread removed. It shows full-length riser cleats secured to both stringer and riser. A division bracket of ¾" plywood is secured between risers and under the treads with glue blocks. **Figure 6** shows the riser mitered to the stringer, and further secured with the full-length cleats glued to both riser and stringer. **Figure 7** is an elevation of the deep riser showing it tapered in width at its length to the bottom of both inside and outside stringers. This tapering of all risers will have no effect on the warped curve of the soffit. **Figure 8** shows a perspective view of a winder tread with two deep risers.

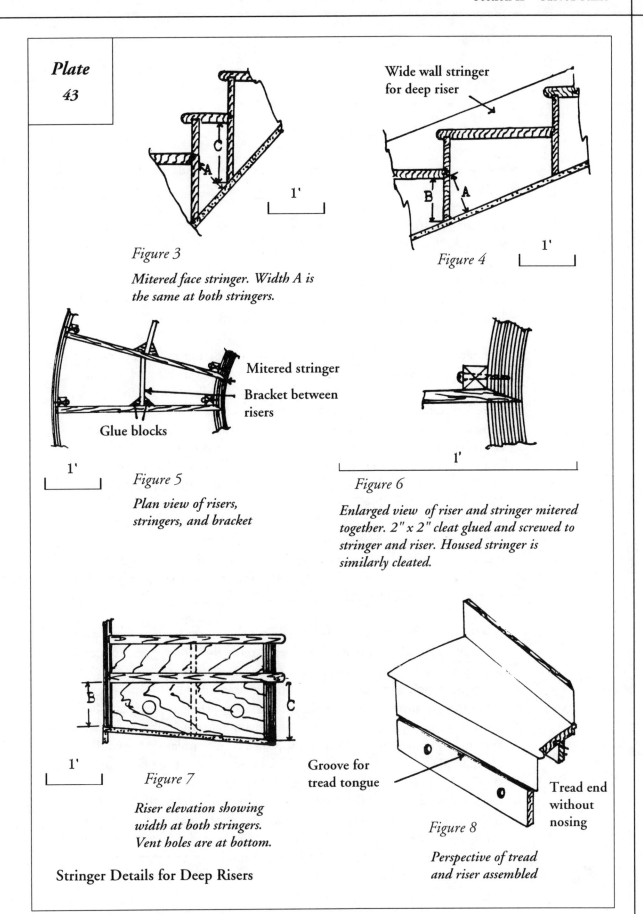

Plate 43

Figure 3

Mitered face stringer. Width A is the same at both stringers.

Wide wall stringer for deep riser

Figure 4

Mitered stringer

Bracket between risers

Glue blocks

Figure 5

Plan view of risers, stringers, and bracket

Figure 6

Enlarged view of riser and stringer mitered together. 2" x 2" cleat glued and screwed to stringer and riser. Housed stringer is similarly cleated.

Figure 7

Riser elevation showing width at both stringers. Vent holes are at bottom.

Stringer Details for Deep Risers

Groove for tread tongue

Tread end without nosing

Figure 8

Perspective of tread and riser assembled

Plates 44 through 45—Making Curved Staircases

Following are progressive steps for making curved or circular staircases such as in **Plate 42.**

1. First make a scale drawing say ¾" or 1" equal to 1' of the proposed stair within the stairwell dimensions.

2. Once the drawing is completed to your satisfaction, make a full-size layout. This is preferably done on drawing paper at a large drawing board, but may also be done on plywood on the floor.

3. Templates for building forms for the laminated stringers for the stair in **Plate 42** are then made. The template radii for both inside and outside stringers should be reduced either $^{3}/_{16}$" or ¼" to allow for a veneer to be placed around the form studs, the convex side representing the face of the laminated stringers. This will ensure a smooth stringer face when the laminations are clamped to the form. Ceiling height permitting, the forms will be vertical rather than horizontal so as to occupy less floor area. The form templates can be made from either 1x12 surfaced stock or ⅝" or ¾" plywood. The template for the inside, or steeper pitched stringer, should extend one tread beyond both lower and upper riser to assure continuity of the stringer curve. This stringer is generally made in one section, whereas the outside, or low pitched stringer, being much longer, is made in two sections. At both forms the stringer face is applied over the form piece veneer.

4. In **Figure 3** of **Plate 44** a hole-boring jig is shown for boring dowel pin holes at both ends of the form studs and in the form templates. Once the form templates are cut to the proper curves they are placed on the full-size plan layout, and the riser positions are marked at the form's edge and along the top surface. Removed from the layout, the templates are then bored for the dowel pins with the hole-boring jig centered at each riser and for stud positions between risers. Once each template is bored, they are then placed over positioned 2x12 common stock at the floor, as shown in **Figure 1, Plate 44** for the inside stringer. With the 2x12 secured to the floor, and the bored template tacked on top, the dowel pin holes are bored into the 2x12 about two-thirds of the thickness. The hole-boring jig is then used to bore the dowel pin holes in the stud ends. The bored form studs can then be used over and over again for many curved forms without destroying the ends with unnecessary nail holes.

**Plate
44**

Making Curved Staircases

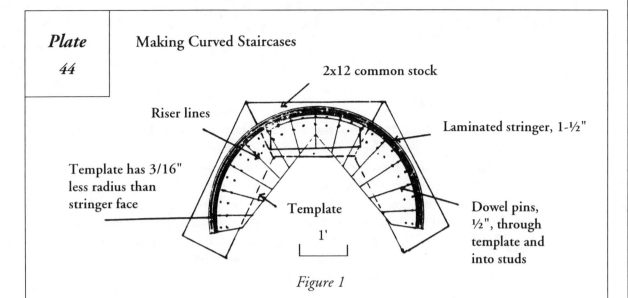

2x12 common stock

Riser lines

Template has 3/16"
less radius than
stringer face

Laminated stringer, 1-½"

Template

1'

Dowel pins,
½", through
template and
into studs

Figure 1

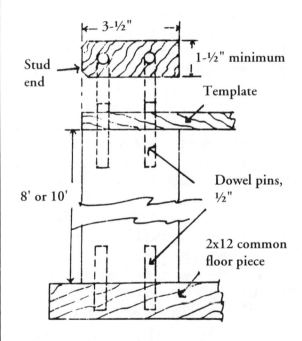

3-½"

Stud
end

1-½" minimum

Template

8' or 10'

Dowel pins,
½"

2x12 common
floor piece

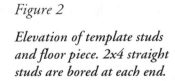

Figure 2

*Elevation of template studs
and floor piece. 2x4 straight
studs are bored at each end.*

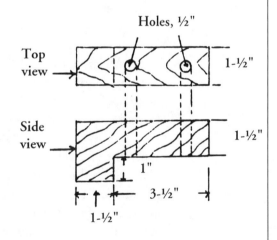

Holes, ½"

Top
view

1-½"

Side
view

1-½"

1"

3-½"

1-½"

Figure 3

*Hole-boring jig for
both template and
studs.*

5. **Figure 2** of **Plate 44** illustrates the 2x4 form studs, between the form template and the 2x12 floor piece, secured in position by the dowel pins. Note that the form face of the 2x4 studs are chamfered at both front edges to make a smooth transition for the form piece from one stud to another. Eight-foot stud lengths are normally adequate for the two-section outside, low pitched stringer, whereas 10' minimum length studs are required for the steeper-pitched stringer. The spacing between studs should be so that there will be no noticeable kinking of the laminated veneers being clamped with moderate pressure. Generally, 5" center spacing is adequate for the low-pitched stringer, and closer spacing for the steeper-pitched inside stringer.

6. **Plate 45** illustrates the template for the low-pitched stringer in **Figure 1** to be comprised of seven risers, or six treads, with a half-lap joint at riser 7, necessitating a continuance of the template curve below riser 7 and above riser 13. **Figure 1** shows the front and side elevation of the form. The form is plumbed and braced with wooden rods to either the floor, wall, or ceiling.

7. The mitered-face stringer is to be 1-½" thick, and the wall stringer, 1" thick. All laminations will be 3/16", plus or minus. The face piece of each laminated stringer is to be clear material of the desired wood, whereas the balance of the laminations can be of common stock. The exposed edge of the stringer easily can be edged with clear material if necessary.

8. Once the face veneers have been laid out for all stringer sections, they are placed atop their respective laminations, including the form piece, on a flat surface. The two stacks of 1" stringer laminations are bored through at the top of risers 4 and 10 with a drill bit just large enough to receive the head of a 10d finishing nail (See **Plate 45**). Likewise, the 1-½" stack of laminated stringer stock, plus the form piece, is bored through at the top of riser 7. Needless to say, whenever the entire width of a stringer is to be exposed, as in a buttress type, this aligning hole should not be bored through the exposed veneer.

9. **Figure 1** of **Plate 45** shows the wall stringer template over the 2x12 floor stock. **Figure 2** shows the stretch out of the erected form. Although not shown here, the inside form is similarly done. Once both forms are erected, draw a plumb line along the stud face of riser 7 at the face stringer form and at risers 4 and 10 at the wall stringer form. Place only the form and face veneers of each stringer on the forms at their approximate pitch so that the nail holes line up with their respective plumb line

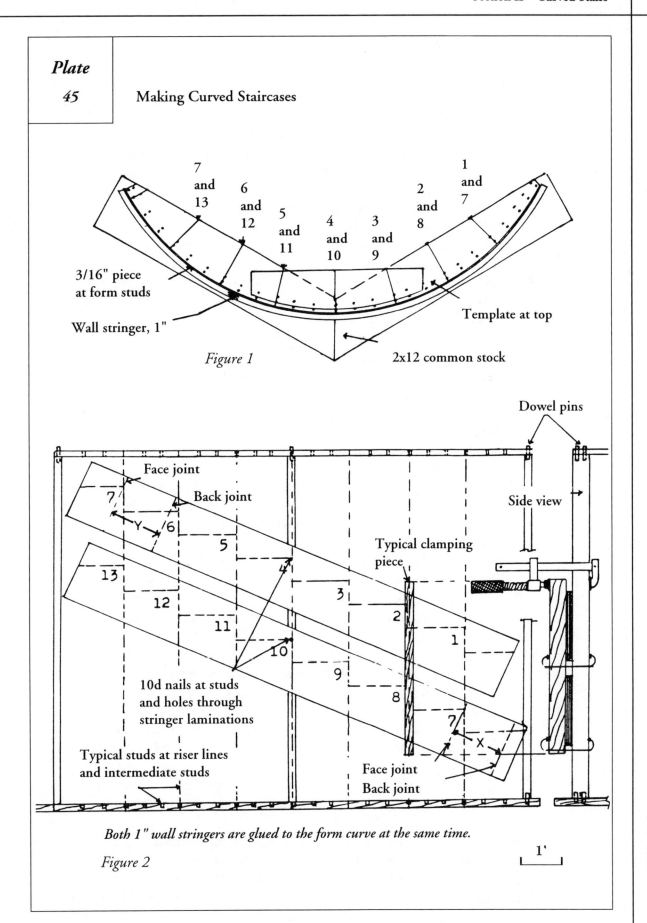

Plate

45   Making Curved Staircases

7
and
13

6
and
12

5
and
11

4
and
10

3
and
9

2
and
8

1
and
7

3/16" piece
at form studs

Wall stringer, 1"

Template at top

Figure 1

2x12 common stock

Dowel pins

Face joint

Back joint

7

Y

6

5

13

12

11

4

3

2

1

10

9

8

Side view

Typical clamping
piece

10d nails at studs
and holes through
stringer laminations

Typical studs at riser lines
and intermediate studs

7

X

Face joint
Back joint

Both 1" wall stringers are glued to the form curve at the same time.

Figure 2

1'

risers. Insert the 10d nails through the pre-drilled holes in the veneers and drive them into the studs at the plumb lines, leaving approximately 1-¾" protruding. Now pivot the two veneers on the nails at the proper pitch so as to attain the correct height from lower to upper riser. Then secure the veneers to the form studs with small nails.

With the face and form veneers now secured to the form, the balance of the veneers are then placed over the protruding nail and clamped at several places to be sure they are aligned with the secured veneers. With half of the stringer length dry-clamped to the form, peel back the laminations of the remaining half and apply the glue to each veneer. Clamp the glued veneers to the form and repeat the process with the other half. In **Plate 45, Figure 2**, note that the half-lap joint of the wall stringer is at riser 7. With this in mind, only glue the back half at Y and the front half at X. Glue the half-laps together once the stringers have been finished.

10. With the glued mitered-face stringer cured, remove it from the form and cut the risers approximately 1" forward of the marked riser line down to the tread line. Then cut the tread line exactly on the line to the riser and remove the loose piece. With a bevel square, take the miter bevel of each riser at the full-size plan, transfer it to its respective tread at the cut stringer, and cut the riser to the bevel thus marked.

11. Now make the wide or "deep risers". See **Plates 42** and **43** for the riser detail and its application for a self-supporting stair. The risers will lock the entire stair into a one piece unit, and at the same time, will form a natural warped soffit line for attaching wire mesh for plastering.

12. The winding treads, starting bull-nosed tread, wedges, cleats, division brackets, and glue blocks can now be made. The stair parts can either be sent to the job site for assembly at the stairwell, or they can be put together in the shop and then sent to the job as a unit. To assist in the stair installation at the job site, a rod, marked from starting to landing floor showing each riser, is used. It is referred to as a "story rod." Since the handrail is made either through laminating at the form or through the tangent principle, the presence of the stair at the shop is not required.

At this point, let us look at laminating a handrail for a curved stair. A laminated handrail consists of gluing enough thin pieces the thickness of the handrail, together around a curved form so as to make the rail width. Once the glue is cured and the rail released from the form, there will be a tendency

Top: Upright forms, often called "drums" or "barrels," are for laminating stair stringers, or, in this instance, wall-type handrails. Form studs are end-bored for dowel pins, so they may be reused in many different configurations.

Left: Curved form templates for a circular staircase show dowel pin holes for each form stud.

Right: Workman clamps handrail laminations to the 2x4s of the barrel.

for the rail to spring to a greater curve than that of the form. This is because the wood fibers of each piece attempt to return to their original shape. Reducing the amount of wood fiber between glue lines by using more and thinner laminations, the springing tendency can be reduced, but not eliminated. If the handrail is glued to the form in its "squared" shape in the above manner, and then molded to the rail profile, the rail will spring even more. If each of the laminations were pre-molded to the rail profile and then glued to the form, there would still be a slight springing tendency. However, the springing may be minimal and the rail could be satisfactory as is. Otherwise, a joint could be made at some point within the section to correct the curve.

When a long freestanding laminated handrail for the entire stair springs to a greater curve than desired, the rail can be cut in one or more places to make butt joints. The joints must then be dressed at the inside in order to return the rail to its proper curve to suit the stair. However, in doing so, proper care must be taken so that there will be no kinking at the joints. I do not have knowledge of any formulas that will take into account wood type, glue type, and the number of laminations used in determining just how much a form curve must be reduced to allow for the molded laminated handrail to spring to its proper curve. The makers of stock circular staircases must adjust their forms to allow for this springing problem.

Unless all laminations are sawn from the same plank and glued together in proper sequence, the grain of a hardwood laminated handrail will not be featured.

As already stated, the lamination of a handrail is restricted to specific conditions, namely, long sweeping curves. Lamination cannot solve most handrail problems as does the tangent system using solid wood stock where the natural wood grain is featured. Laminating does, however, enable anyone familiar with woodworking to make a suitable curved stair to accommodate stock handrail parts.

Plate 46—The Stretch-Out of the Mitered-Face Stringer of Plate 38 and Kerfing and Staving Short-Radius Stringers

Figure 1 shows the stretch out of the face of the curved stringer. It can either be laid out on a strip of drawing paper or on the veneer itself.

Figure 2 is a detail showing how short-radius stringers are either staved or

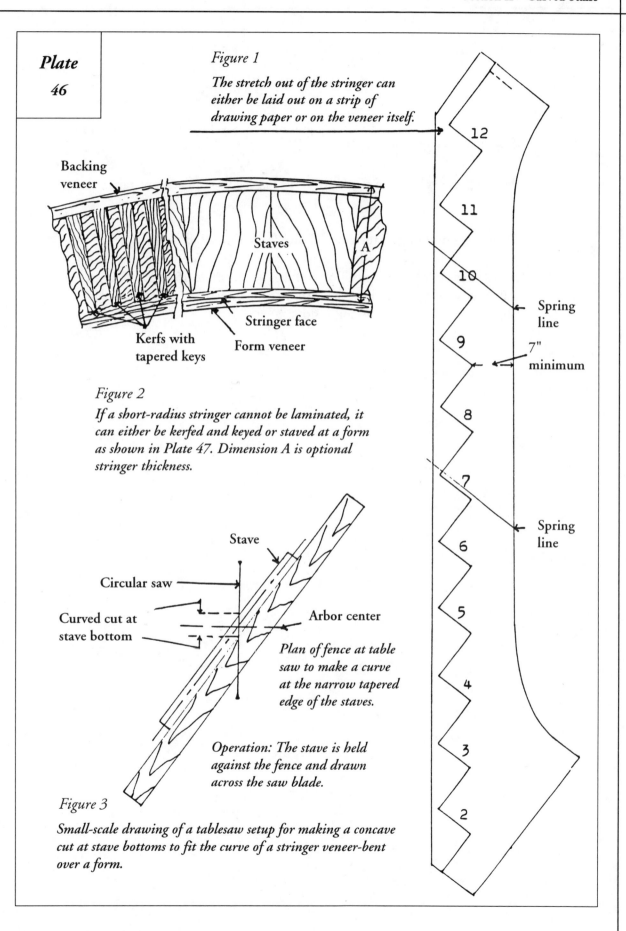

Plate 46

Figure 1

The stretch out of the stringer can either be laid out on a strip of drawing paper or on the veneer itself.

Backing veneer

Staves

A

Stringer face

Form veneer

Kerfs with tapered keys

Figure 2

If a short-radius stringer cannot be laminated, it can either be kerfed and keyed or staved at a form as shown in Plate 47. Dimension A is optional stringer thickness.

Stave

Circular saw

Curved cut at stave bottom

Arbor center

Plan of fence at table saw to make a curve at the narrow tapered edge of the staves.

Operation: The stave is held against the fence and drawn across the saw blade.

Figure 3

Small-scale drawing of a tablesaw setup for making a concave cut at stave bottoms to fit the curve of a stringer veneer-bent over a form.

12

11

10

Spring line

7" minimum

9

8

7

Spring line

6

5

4

3

2

kerfed and keyed instead of being laminated with thin members. The dimension shown at A is optional.

Figure 3 shows how the staves are curved at the bottom at a saw table.

Plate 47—Building a Horizontal Form for a Short Radius Stringer

Whenever building a vertical form of 2x4's as in **Plate 45** becomes impractical because of a tight radius, the form may be built as shown in this plate. **Figures 1** and **2** show how this is done.

Plate 47

Building a Horizontal Form for Short Radius Stringers

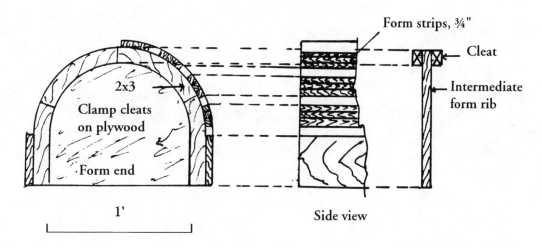

Form strips, ¾"

Cleat

Intermediate form rib

2x3

Clamp cleats on plywood

Form end

Side view

1'

Figure 2

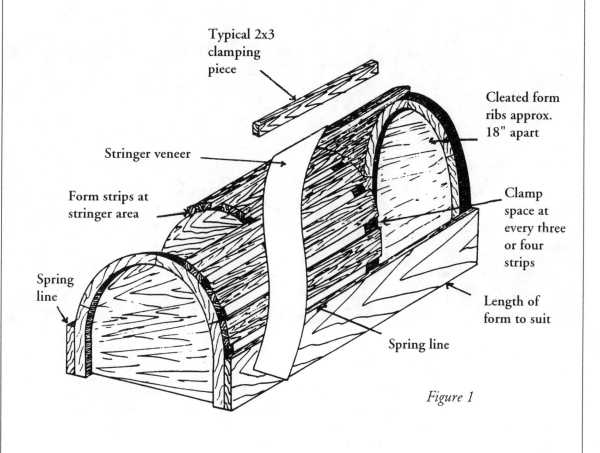

Typical 2x3 clamping piece

Stringer veneer

Form strips at stringer area

Spring line

Cleated form ribs approx. 18" apart

Clamp space at every three or four strips

Length of form to suit

Spring line

Figure 1

This curved stair of all red oak features an exposed soffit with an open-side, mitered face stringer. The wall stringer is 1" thick housed and wedged, and the mitered-face stringers are 1-1/2" thick laminated Douglas fir with 3/16" red oak veneer. Deep risers are 1" thick red oak to the bottom of the treads, with Douglas fir extended to the bottom of the stringers. The deep risers both strengthen the stair and provide a natural warped soffit to receive the plaster. The mitered face stringer at the open side eases to the wide, curved balcony fascia.

The installation uses stock handrail volutes, easements, posts, and balusters; the curved railing is custom laminated. The interior view, above, shows the warped soffit easing to the balcony ceiling, and the handrail curving around the stairwell balcony. This residence is in Modesto, California.

Section III

Tangent Handrailing

Making both straight and curved staircases has been adequately covered in **Sections I** and **II**. The learner should have no difficulty in understanding the methods for designing such stairs. I have already mentioned at the beginning of this book the importance of having a basic knowledge of tangent handrailing in designing a stair requiring a continuous handrail. Up to this point I have covered only the essentials of tangent handrailing.

This section encompasses the entire field of tangent handrailing from the most complex of twisted molded-type handrail turns, to the simplest full-round type incline-turns. It demonstrates the use of geometrical drawings to produce the required face mold and bevels necessary in order to make incline-turn solid wood handrail sections for any possible handrail condition. The method is not difficult to follow. It is progressive and precise.

Other methods exist, of course, with the practice of building forms to laminate, and the time-consuming, trial-and-error fitting of blocks to suit, among them. However, I know of no other method of handrailing more precise, compact, and comprehensive than the method presented here.

I believe this treatise will simplify understanding the tangent system of handrailing by showing, not only how, but why specific lines are drawn to determine the angle of the elevated tangents, the bevels, and the face mold, all equally important in making a precise handrail section. If the slightest error is made in any of the three, the rail section is useless. In order to prevent such an error, I show several ways to prove each step before proceeding to the next. By doing so, the railmaker is assured of an accurate result.

The principles of tangent handrailing have been graphically shown by the prismatic solid of **Plate 30** and the cardboard model of **Plate 31**. This book has a quick reference guide to every tangent pitch or level combination for any plan type in the chart in **Plate 72**.

I have also included a number of layouts of various types of stairs involving climbing-turn handrail sections and the pitch of the tangents for making their incline-turn handrail sections. If you are currently using another method to produce incline-turn handrail sections, you may find this method to provide helpful shortcuts, improvements, or supplemental information to your present system.

The learner may wish to analyze the tangent system shown here and study the explanations. However, the method is presented in such step-by-step fashion that, as long as you follow the progression, you can be confident of precise results.

Plates 48 through 52—The Basic Plan Tangents of Plate 3 Shown as Incline-Turn Tangents of a Handrail Section

The next five plates show the three basic plan curves of **Plate 3** as centerline curves of incline-turn tangent handrail sections. The two tangents of each section will be either pitched equally or unequally, or one tangent will be pitched and the other level. The tangents are to incline from points A to C in **Figures 1** through **15**. The elevation (height through the turn) is shown as GH. Only the inside curve of half the plan rail width is shown. In all radius drawn plans, the plan tangents are shown to be equal length. In subsequent elliptical plans, one plan tangent is shorter than the other.

In all 15 of the following figures, unfolding or stretching out the plan tangents is the first step toward finding the shape termed the face mold pattern in order to make the incline-turn section at hand. The face mold will be further explored in subsequent plates. The lower tangent is stretched out along the line of the upper plan tangent as BD. The spring lines at C and D, as well as the vertex line, are extended vertically. The tangents are then drawn pitched or level as they might occur over the risers of a stair and at the starting volute or landing levels.

The horizontal base line is a level line between the vertical extension of the stretch out of the low point of the lower tangent at D, and the vertical extension of the uppermost point of the upper tangent at C. These points are generally, but not necessarily, the spring lines of the plan curve. The horizontal base line originates from where the pitch of the lower tangent crosses the lower vertical extension at E. **Plates 48** through **52** show the horizontal base line as EG, with GH the height through the turn. At each of the 15 figures, there is also an isometric view of a simulated prismatic solid with the pitched tangents standing over the plan tangents, showing the oblique plane formed by the tangents and their parallels. The elevated tangents are shown as t. These prismatic views show that whenever the tangents are unequal, the oblique plane not only slopes in length, but also in width, as shown in all figures except at 1-A, 2-A, and 3-A. The height of the oblique plane opposite the vertex of the tangents is the difference between the total height and that at the vertex. The inside curve of the handrail width at the point of intersection of the plan tangents' parallels is extended vertically to intersect the parallels of the pitched tangents at the oblique plane as at J and K.

In the obtuse and acute plans, the outside curve of the rail width at the plan tangents' parallels may also have to be extended to the oblique plane. This procedure is further explained in the ordinate and face mold, **Plate 58**, for finding the widths of the face mold along the oblique plane parallels at the spring-line points.

These combinations of plan tangent angles and tangent pitch or level combinations from **Plates 48** through **52** comprise all of the possible tangent conditions in handrailing. Solving the elevation and pitches of the tangents is an essential step toward finding the angle of the tangents and the face mold.

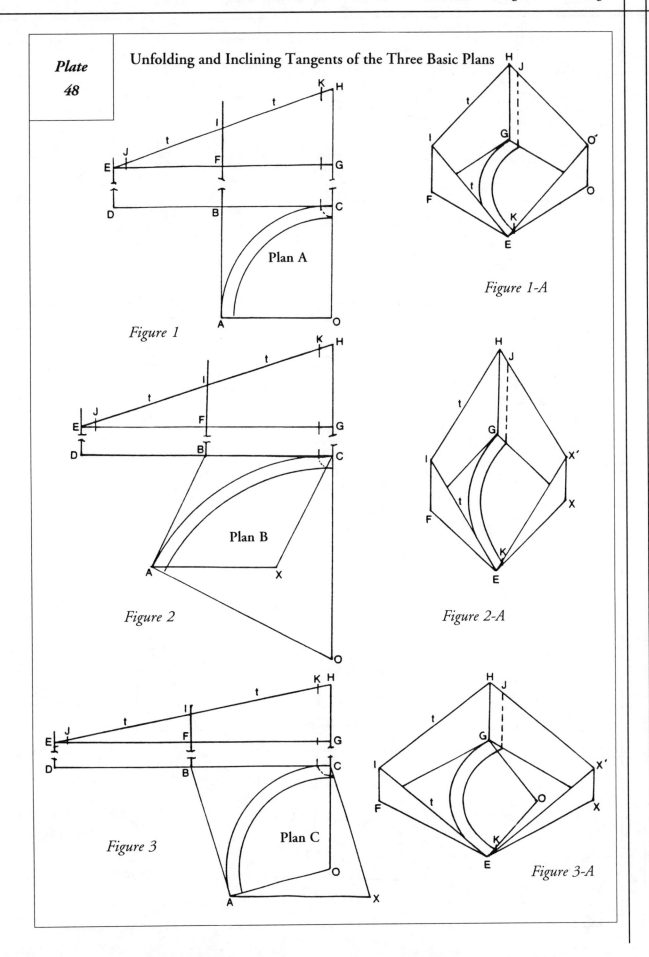

Unfolding and Inclining Tangents of the Three Basic Plans

Plan A

Figure 1

Figure 1-A

Plan B

Figure 2

Figure 2-A

Plan C

Figure 3

Figure 3-A

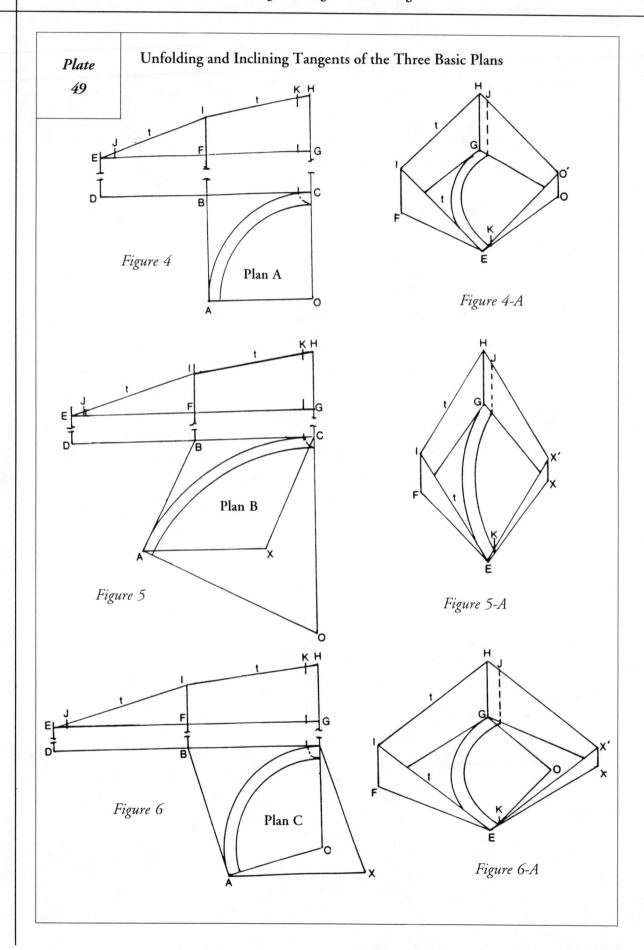

Plate 49

Unfolding and Inclining Tangents of the Three Basic Plans

Figure 4

Plan A

Figure 4-A

Figure 5

Plan B

Figure 5-A

Figure 6

Plan C

Figure 6-A

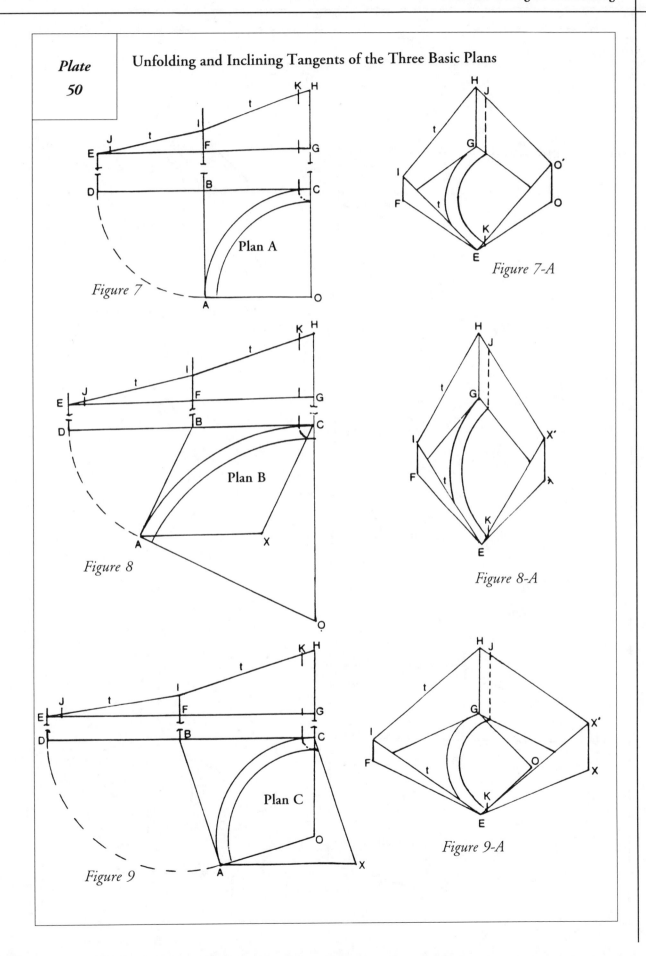

Plate 50

Unfolding and Inclining Tangents of the Three Basic Plans

Plan A

Figure 7

Figure 7-A

Plan B

Figure 8

Figure 8-A

Plan C

Figure 9

Figure 9-A

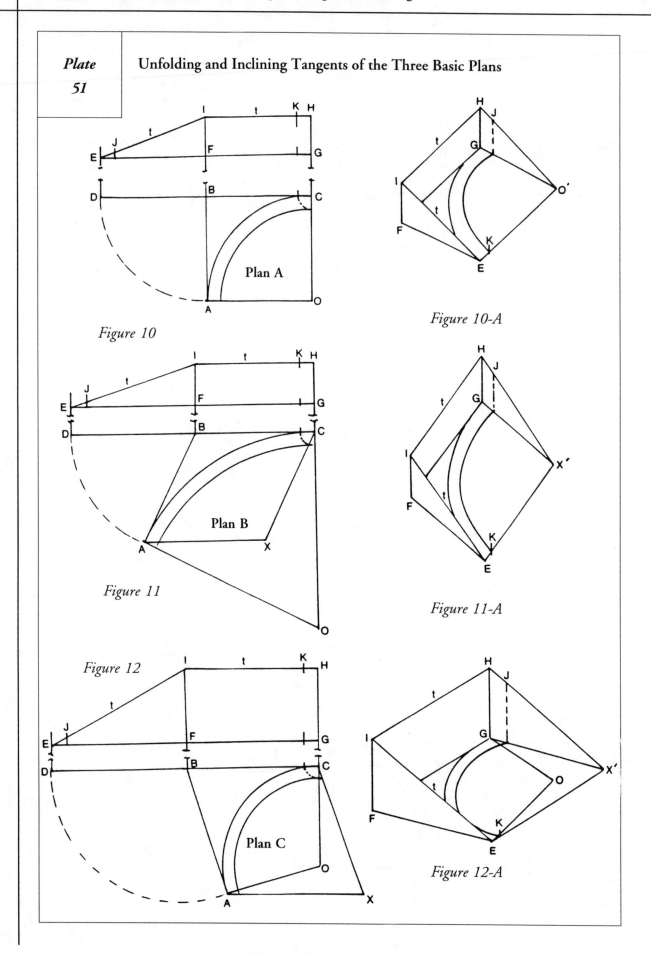

Plate 51   Unfolding and Inclining Tangents of the Three Basic Plans

Figure 10

Plan A

Figure 10-A

Figure 11

Plan B

Figure 11-A

Figure 12

Plan C

Figure 12-A

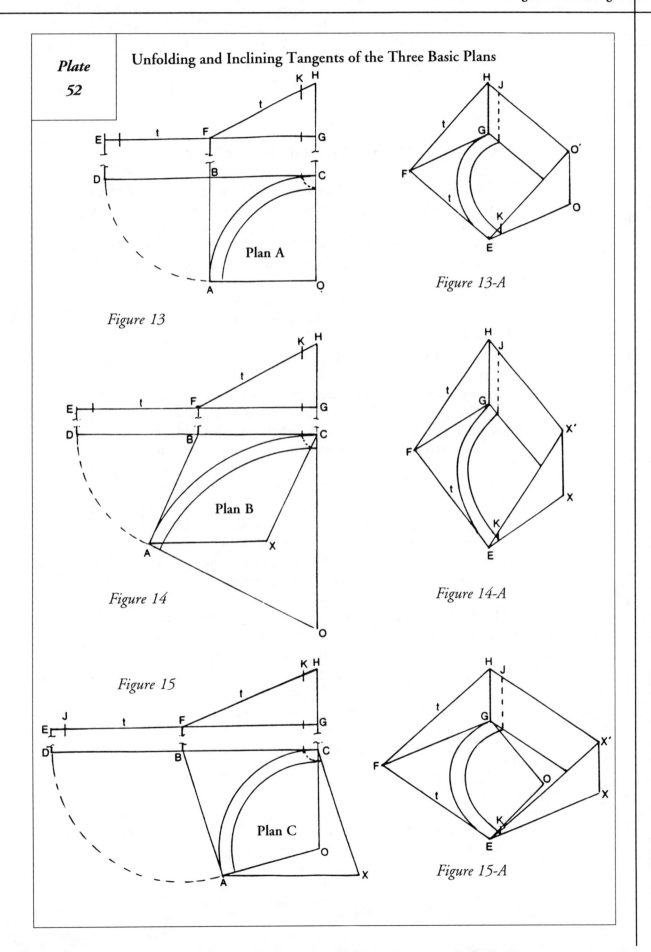

Plate 52

Unfolding and Inclining Tangents of the Three Basic Plans

Figure 13

Plan A

Figure 13-A

Figure 14

Plan B

Figure 14-A

Figure 15

Plan C

Figure 15-A

Plates 53 through 57—The Angle of the Inclined Tangents

Once the tangent pitches have been determined and the horizontal base line drawn, the inclined tangent angles can then be found as they would stand over their respective plan tangents. The following figures of all of the possible plan types and tangent pitch or level combinations in handrailing show how this angle is obtained in three manners at the pitched tangents.

1. With the pitched tangents drawn above the plan in all figures, strike an arc with CA as the radius to intersect the horizontal base line at I. With DI as the radius, strike the arc IG, extended. Now strike an arc with the lower-pitched tangent EF in **Figures 1** through **12** and level tangents BF in **Figures 13** through **15** as the radius to intersect DI at G. Connect GE to make the angle of the tangents as shown in **Figures 1** through **12**, and connect GB in **Figures 13** through **15** for the angle. This method, while accurate, is not recommended as standard practice for extremely low-pitched obtuse plans because it is difficult to determine the exact point where two arcs cross.

2. Draw a perpendicular line from the horizontal base line to the low point of the lower plan tangent at A as line AH; at right angle plans it is always AB. From point B in right angle plans and from point H in obtuse and acute plans, draw a perpendicular line to the upper tangent to intersect the arc of the lower tangent at G. Connect GE in **Figures 1** through **12** and GB in **Figures 13** through **15** to form the angle as shown.

3. The third method, not shown here, is found by the ordinate direction line and the minor axis direction line being equal lengths from the horizontal base line as in **Plates 58** through **62**.

Plate 53

The Angle of the Inclined Tangents

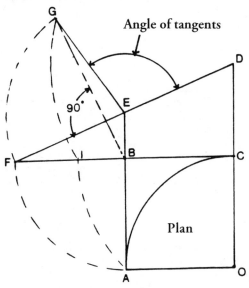

Figure 1

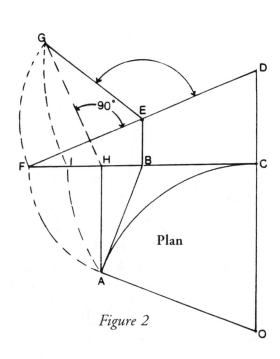

Figure 2

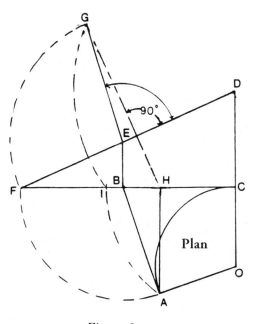

Figure 3

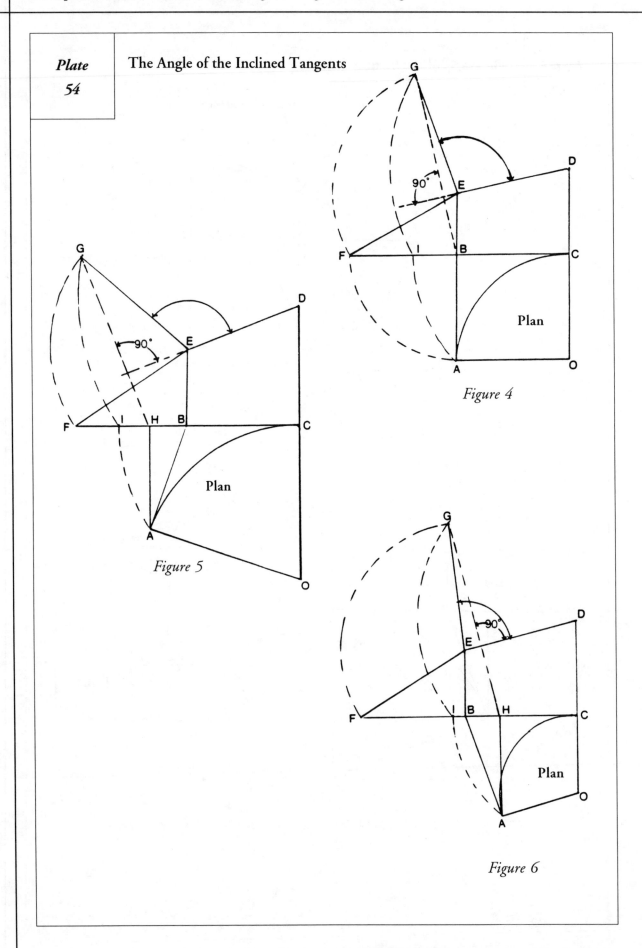

Plate 54

The Angle of the Inclined Tangents

Figure 4

Figure 5

Figure 6

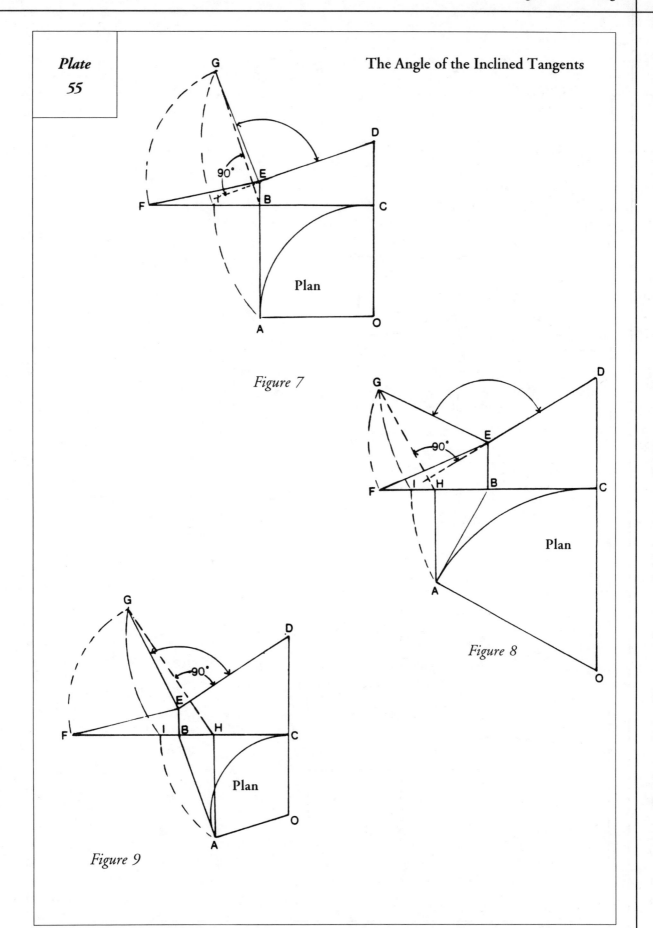

Plate 55

The Angle of the Inclined Tangents

Figure 7

Figure 8

Figure 9

Plate 56   The Angle of the Inclined Tangents

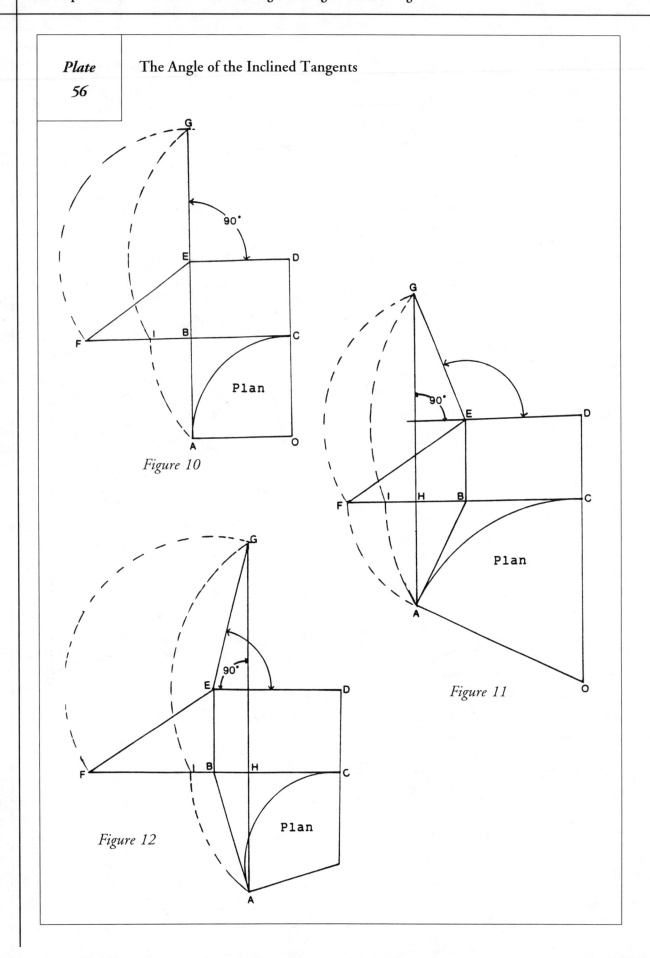

Figure 10

Figure 11

Figure 12

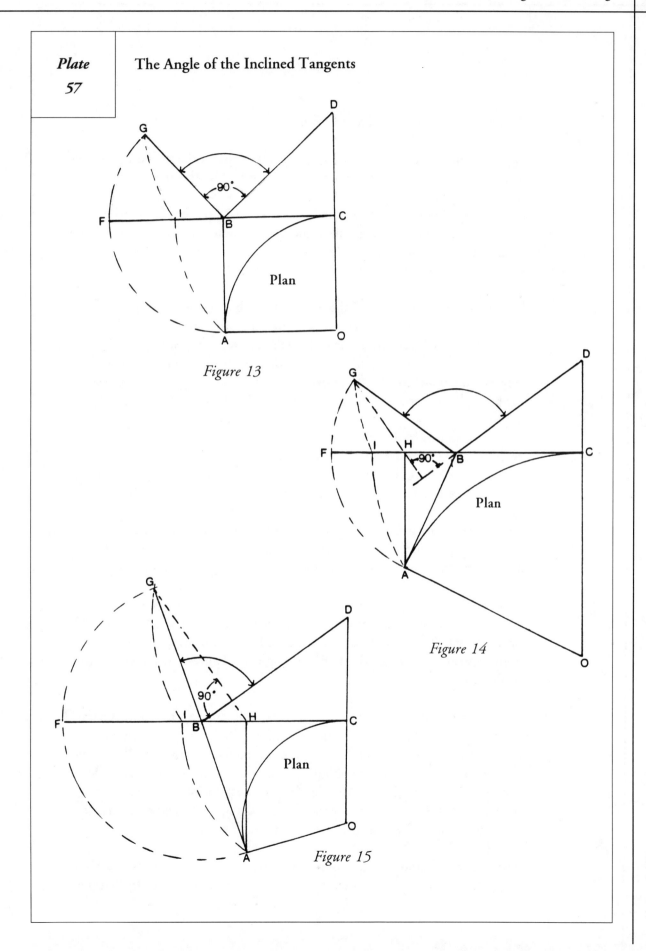

Plate
57

The Angle of the Inclined Tangents

90°

Plan

Figure 13

90°

Plan

Figure 14

90°

Plan

Figure 15

Plate 58—The Face Mold and Ordinate

When the inside and outside curves of the rail width at the plan are projected to show on the surface of the oblique plane formed by the angle of the elevated tangents, the elliptical pattern of the rail width on the plane is termed the face mold. The face mold is the essential pattern required in order to make an incline-turn handrail in the tangent system of handrailing. Joints of the face mold are made perpendicular to the tangent lines marked on the surface of the face mold.

In order to draw the shape of the face mold at the angle of the elevated tangents, the direction of a level axis termed the ordinate must first be found at the plan. The direction of the ordinate's counterpart, the minor axis, is then found at the oblique plane. Both ordinate and minor axis are equal length and parallel to each other over the plan. While the ordinate is level at the plan, the minor axis lies level along the oblique plane. The ordinate originates from the radius point of a radius drawn plan. The direction in which it is drawn from the radius point with respect to the plan tangents is determined by an ordinate direction line. The ordinate direction line is found by extending the pitch of the upper tangent to strike the horizontal base line (see **Plates 48** through **52**), plumbed to the stretch out line of the plan tangents (which may also be the horizontal base line) and then connected to the low point of the lower tangent. The minor axis direction lines at both figures equal their respective ordinate direction lines in order for the angle of the elevated tangents and the shape of the face mold to be correct.

Finding the angle of the elevated tangents and the shape of the face mold, using the pitch of the upper tangent, will be referred to as the **Method "A"** principle, which will be the primary method used in this book. It is shown in both **Figures 1** and **2. Method "B"** principle, in **Figure 2**, will be explained further along.

With the ordinate direction line drawn, the ordinate is simply drawn parallel from the radius point, as at OB in both figures. To find the face mold curves through either the trammel or string and pin methods shown in **Plate 2**, the ordinate is transferred to the oblique plane of the elevated tangents as the minor axis B'O' in **Figure 1**. In **Figure 1**, at the plan, KL is drawn perpendicular to BO. At the oblique plane, ST is the major axis perpendicular to B'O'. A'Q and C'R equal AM and CN at the plan.

The trammel method is used to find the face mold, or for finding pin placements for the string and pin method of drawing the face mold. In **Plates 48** through **52,** the prismatic views of the oblique planes show how the widths of the face mold joints are found at the parallels to the elevated tangents as they stand over the plan. In the right-angle plan of **Plate 58, Figure 1,** CG or D' is half the plan rail width. Extend I to J and G to H. A'3, along the parallel A'O', is marked as DJ. Likewise, C'8, along parallel C'O', is marked as C'H. In some obtuse and acute angle plans CG may not be the same margin at both sides of the spring line along the plan tangents' parallels. If the margins are different then both margins must be transferred to the elevated tangents' parallels.

This procedure can be done when using the trammel or string and pin method. At the minor axis B'O', let B'-1'-2' equal B-1-2 at the plan. To find the trammel lengths for the inside curve of the face mold, let 3-4 equal 2'-0', and extend to 5 along the minor axis line B'O'. With points 3, 4, and 5 marked on a rod of suitable length, move the rod with point 4 kept along the major axis, and point 5 along the minor axis line, marking point 3 at various places through which the inside curve can be drawn from A' to C'. The outside curve is found in the same manner. 8-6 equals 1'-0', and is extended to the minor axis line at 7. Point 6 is kept along the major axis, while 7 is kept along the minor axis line, marking point 8 at various places from A' to C' through which the outside curve is drawn.

To draw the same face mold using the string and pin method of **Plate 2,** trammel lengths 3-5 and 7-8 are used to find the pin placements along the major axis. Let 2'-p1 equal 3-5 for pin locations for the inside curve. Let 2-p2 equal 7-8 for pin locations for the outside curve. With a taut linen line secured to the proper pin for the curve to be drawn, looped around the other pin, and stretched to the rail width at the minor axis, the face mold curves are drawn. See **Plate 2.**

An optional method of finding the pin placements for drawing the face mold with either the trammel or string and pin method is to extend the plan curves of the rail width to the diameter line KL at U and V. U and V are then transferred to the major axis ST at W and X as shown. O'W equals 2'-p1, and O'X equals 1'-p2.

Figure 2 is the same plan and tangent pitch combination as in **Figure 1** except the face mold is found using either the Method "A" principle or the equally accurate Method "B" principle. Both face molds are found using parallels to the plan ordinate or ordinate direction. In either method, random parallels to the ordinate or ordinate direction are drawn as 1 and 2 at the plan to intersect the plan tangents. In Method "A", they are transferred to the stretch out of the plan tangents, to the pitched tangents, and then to the angle of the elevated tangents, from where they are drawn parallel to the minor axis or minor axis direction the same length from the elevated tangents as they are from the plan tangents to establish points through which the face mold curves can be drawn. For all practical purposes, unless the trammel or string and pin methods are to be used to draw the face mold, the width of the face mold joints are twice that of JR at the bevel. The bevel will be fully explained in **Plate 64.**

In **Figure 2** to find the angle of the tangents and the face mold using the Method "B" principle, a perpendicular to the ordinate direction FA, called the "seat," is drawn. In equally pitched tangents such as in this plan, the seat may be AC. However, so that the layout will not encroach upon the plan itself, let the seat be drawn parallel to AC through O, as at KL. Set up LM perpendicular to KL, and equal to height CD. Connect MK as the pitch. From vertex B at the plan, draw BP parallel to FA.

Perpendicular to MK, PQ equals NB. Connect QM and QK for the angle of the elevated tangents. Square K and M to their tangents. JR, at the bevel, is half of the face mold joint widths. The same random plan parallels used to find the face mold in Method "A" are now used to find the face mold in this Method "B" principle. Parallels 1 and 2 are extended from the plan to strike the pitch line KM, from which points they are drawn perpendicular to KM the same distance as they are from AC to the plan curves. All points thus marked are connected to form the shape of the face mold.

As stated earlier, the bevel at J is explained in **Plate 64.** The bevel will determine the width and thickness required for the handrail block for the section as shown at Z and Y in **Figure 2.**

Proof of the accuracy of the face mold layout in Method "B", is that if each line, AC, KL, and KM, were an axis by folding along these lines, the curves of the face mold would be shown to lie directly over the plan curves. **Figure 3** shows an elementary but effective method of quickly drawing parallel lines using a weighted straight edge and a triangle with a minimum 3' hypotenuse. The triangle is simply moved along the stationary straight edge to draw parallels as desired.

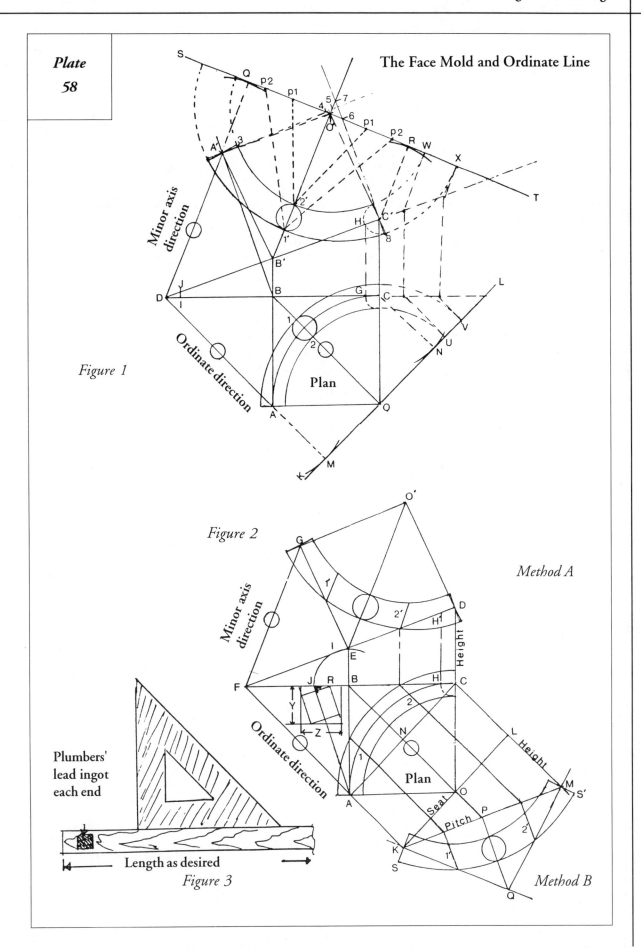

Plate 58

The Face Mold and Ordinate Line

Figure 1

Minor axis direction

Ordinate direction

Plan

Figure 2

Minor axis direction

Ordinate direction

Plan

Method A

Method B

Plumbers' lead ingot each end

Length as desired

Figure 3

Plate 59—Unequal Tangents Affect the Ordinate Direction

This plate is an example of how the pitch of the upper-tangent affects the direction of the ordinate and the minor axis. The extension of tangent C'B' strikes the base line at H. HA is the ordinate direction line. From radius point O, draw O4 parallel to HA. If either the trammel or string and pin method is to be used to draw the face mold, make PQ square to HA through radius point O. Let A'T and C'U at the oblique plane equal AR and CS at the plan. Connect TU as VW. Draw 4'O' parallel to HA'. Connect O'C' and O'A'. C'N and C'N' equals GK, and A'M and A'M' equals C'L, as shown in **Plates 48** through **52**. To draw the face mold through various points, draw random parallels to the ordinate at the plan through the rail width at 1, 2, 3, and 4, and then transfer them to the oblique plane as 1', 2', 3', and 4'. The extreme points, including MM' and NN', are then connected to form the face mold shape, as in **Figure 2** of **Plate 58**.

With the major axis line VW and points N, N', M, and M' established, the face mold can also be drawn by either the trammel or string-and-pin methods shown in **Figure 1, Plate 58**.

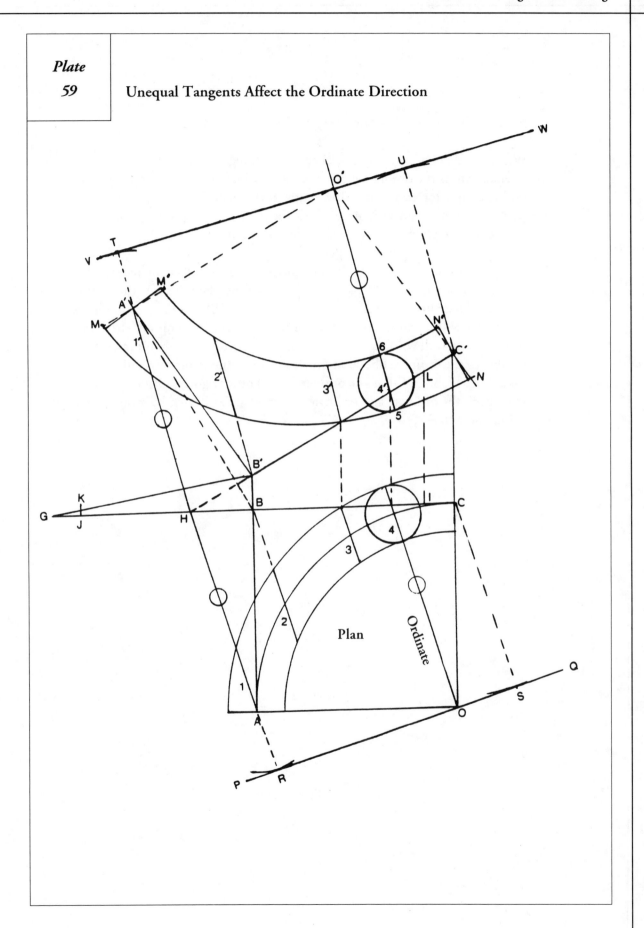

Plate
59

Unequal Tangents Affect the Ordinate Direction

Plan

Ordinate

Plates 60 Through 62—The Ordinate Direction and Ordinate

Whenever there is a level and pitched tangent, regardless of plan type, the level tangent is the ordinate direction. The ordinate is drawn parallel to the ordinate direction from the radius center at the plan. In any plan type, whenever the two tangents are pitched, the pitch of the upper tangent will dictate the direction of the ordinate from the radius center.

The point from where the pitch of the upper tangent strikes the horizontal base line is transferred to the stretch out line of the plan tangents (unless the horizontal base line is already the stretch out line, which is the case in **Figures 1** through **9**) from which point it is connected to the low point of the lower plan tangent as the ordinate direction line. The ordinate is then drawn parallel to the ordinate direction line from the radius center. In following **Figures 1, 4,** and **7**, OB is the ordinate. In all other figures, OQ is the ordinate. Both the ordinate and ordinate direction are indicated by a small circle through the lines.

Plate 60 The Ordinate Direction and Ordinate

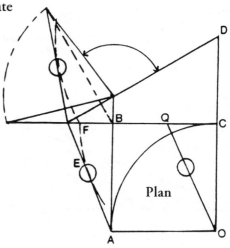

Figure 2

In Figure 2, CE is the distance
between C and ordinate direction,
intersecting the baseline at F. FD is
then the distance from D to the
minor axis direction line.

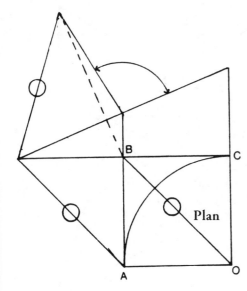

Figure 1

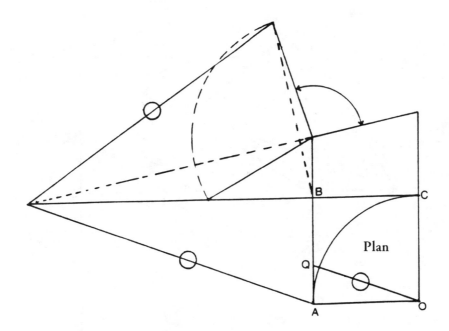

Figure 3

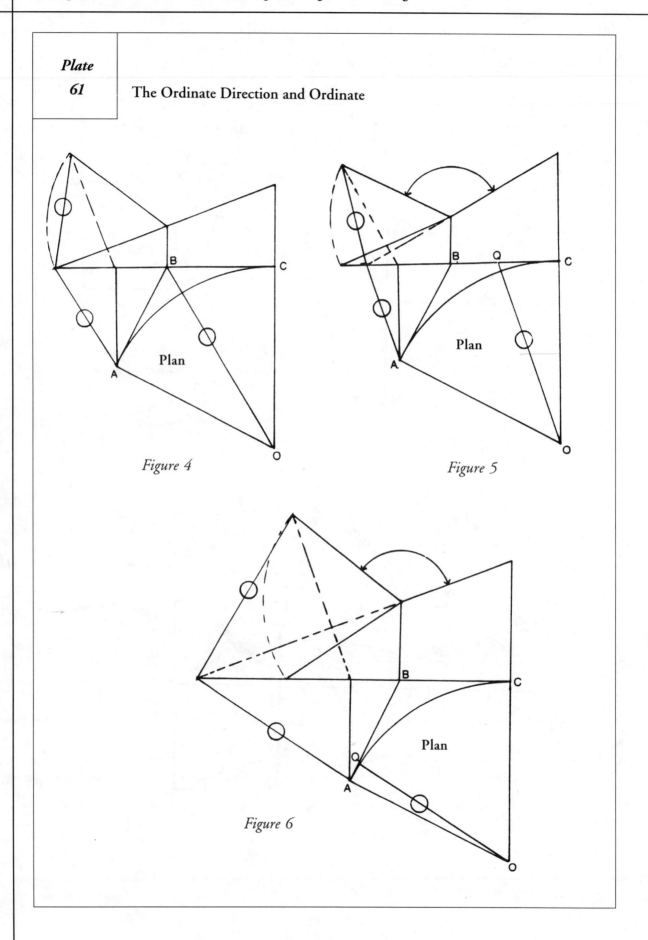

Plate 61   The Ordinate Direction and Ordinate

Figure 4

Figure 5

Figure 6

Plate 62   The Ordinate Direction and Ordinate

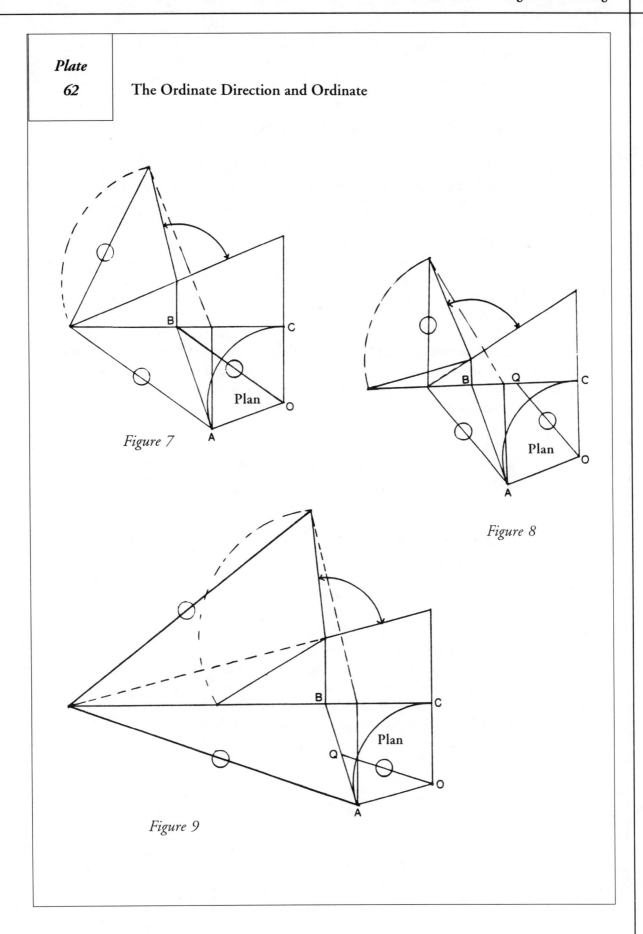

Figure 7

Figure 8

Figure 9

Plate 63—Alternate Methods of Finding Ordinate Direction and Ordinate

Whenever the extension of the upper pitched tangent intersects the horizontal base line beyond reasonable layout area, the ordinate direction line can be found by two alternate methods shown in this plate.

In **Figures 1, 2,** and **3,** let the pitch of the lower tangent be extended to intersect a level line drawn from the height CD.

For the first alternate method, in **Figure 1,** let CI equal DH. Let OJ equal BI. Connect JA for the ordinate direction line. Ordinate OK is drawn parallel to JA, BK is transferred as BL, L is extended to M, EN equals EM, and N is connected to O' at the parallelogram of the angle of the tangents for the minor axis. The major axis is drawn perpendicular at O' if either the trammel or string and pin methods are used to draw the elliptical curves of the face mold.

In **Figures 2** and **3** as in **Figure 1,** CJ equals DI and KL equals BJ along the plan tangent's parallel CK. LA is the ordinate direction line. KN and ordinate OM are transferred to the oblique plane parallelogram as QR and TO'. TO' is the minor axis line, with the major axis at O'.

For the second alternate method, in **Figure 1,** DH is simply marked as BL, with BK equal. Connect KO for the ordinate. Parallel AJ need not be drawn. BL is transferred to the pitched tangent FE and then to N as shown. Since the parallelogram of the elevated tangents establishes O', N is connected to 0' as the minor axis. The perpendicular line drawn at O' is the major axis. For the second alternate method in **Figures 2** and **3,** BP is marked as DI, with BN equal length. BP is transferred to the tangent EG at Q. Q is connected to R of the oblique parallelogram as the minor axis direction. At the plan, N is connected to K of the plan parallelogram for the ordinate direction. Then, OM is drawn parallel to NK as the ordinate. M is transferred to elevated tangent GE at T. TO' is drawn parallel to QR the same length as MO at the plan for the minor axis.

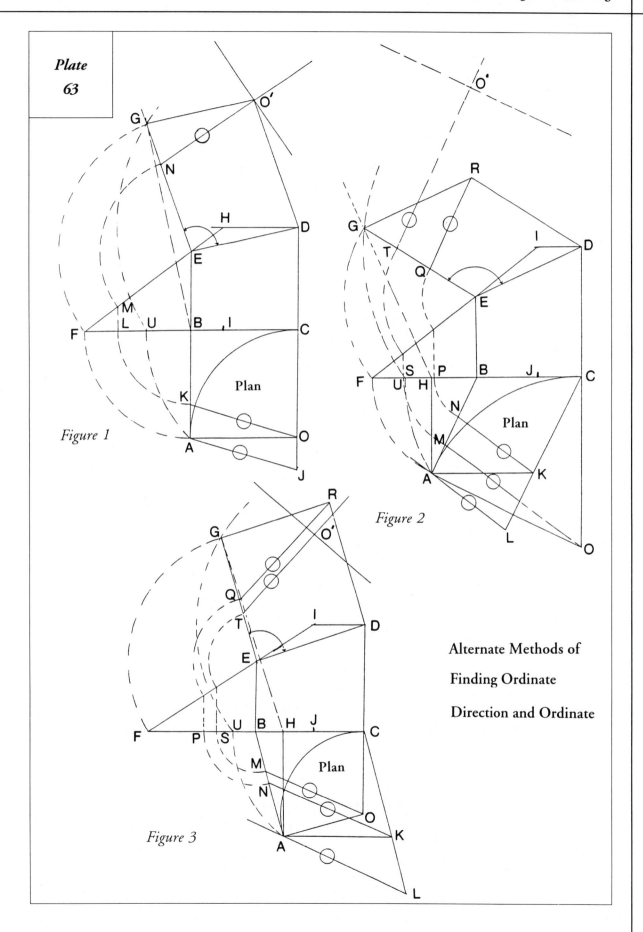

Plate 63

Figure 1

Figure 2

Figure 3

Plan

Plan

Plan

Alternate Methods of
Finding Ordinate
Direction and Ordinate

Plates 64 through 66—The Bevels

Figures 1 through **5** in **Plate 64** show four methods of finding the bevel, or bevels, that must be applied to the butt joints of an incline-turn handrail section in order to impart the twist through the turn.

Figure 1 in **Plate 64** shows a solid square block cut on a diagonal so as to simulate inclined tangents of a quarter-turn handrail section. Assume DE and DA are unequally pitched tangents to the dotted curve shown on the oblique plane, the plan tangents being AB and BC. Bevel squares are shown applied from the surface of the plane to the side of the block form plumb line bevels for joints D and A. These same bevels can be found through a surface layout by four different methods.

First method: **Figure 2, Plate 64**—This will be the primary method of finding bevels for any incline-turn handrail section in this book. It is recommended over all other methods simply because the bevels are quickly and accurately found at the pitch of the tangents with a minimum number of lines. The bevel for joint D is found by striking an arc from B to touch the extended pitch of DE, intersecting the base line at M. M is connected to A (the low point of the lower tangent) for the required bevel at M. To find the bevel for joint A", extend AB to equal height CD as BH.

From H, strike an arc touching the extended pitch A'E, or equal pitch line CE, as $\underline{w}$. Let BL equal $\underline{w}$. Connect LA for the bevel for joint A". Side $\underline{y}$ in **Figure 2**, shows the parallel height $\underline{v}$ between the upper pitched tangent ED and its opposite parallel BO". **Figure 3** shows side $\underline{z}$ of **Figure 2**, with W the same in all **Figures 1, 2,** and **3**.

Second method: **Figure 2, Plate 6 4**—In the oblique plane parallelogram, let X and Y be respective hypotenuses of the right triangles of the bevels at L and M. Since a precise parallelogram must be drawn, especially for low-pitched obtuse plans where a slight error in drawing the oblique parallelogram is possible, this method should only be used as a means to check the accuracy of the other methods.

Third method: **Figure 4, Plate 64**—This is the same plan and tangent pitch combination as at **Figure 1**. Where the bevels in method one are found through the pitch of the tangents, this method requires that the angle of the tangents be found first, as in the method "B" principle of **Plate 58**. At the meeting point of the pitch and height at H, draw HL and HM parallel to respective tangents AB' and B'D'. At any point along the pitch AH, draw a

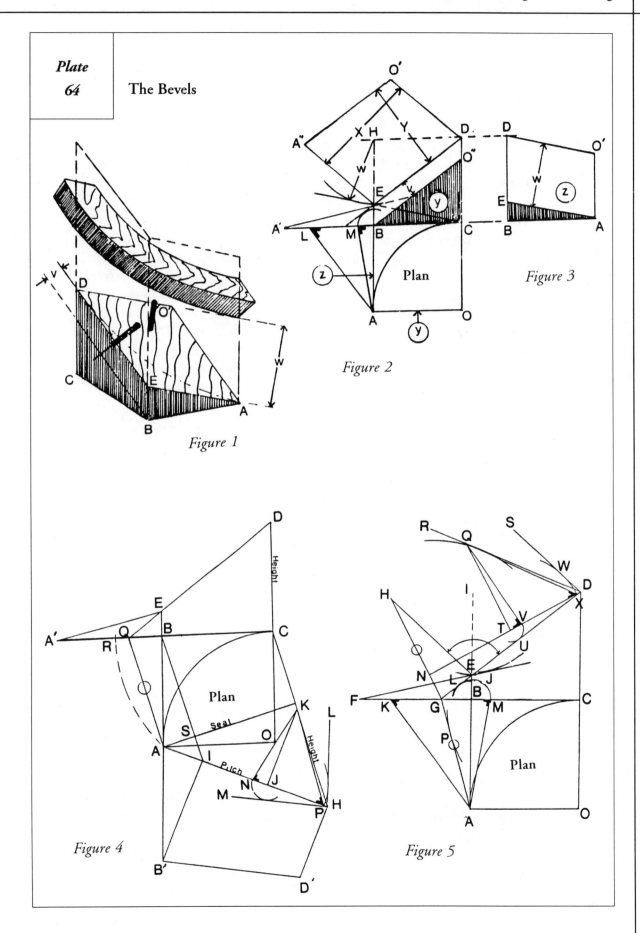

Plate 64 The Bevels

Figure 1

Figure 2

Figure 3

Plan

Figure 4

Plan

Figure 5

Plan

perpendicular line, say JK, to intersect the height line. With J as a radius point, strike an arc touching parallel HL and HM to intersect the pitch line at P and N. Connect PK for the bevel at joint A, and connect NK for the bevel at joint D.

Fourth method: **Figure 5, Plate 64**—Like method one, the angle of the elevated tangents must be found by first using the Method "A" principle of finding the angle of the elevated tangents. In this fourth method, it is critical that the minor axis direction be accurately drawn (see **Figure 2, Plate 60**). From the minor axis direction line GH, draw a perpendicular to the height point D as ND. From N strike an arc equal to arc CP, as NQ. From D draw a line through arc NQ, as DR. Again, from D draw a parallel DS to tangent EH. From any point along ND, say at T, draw a perpendicular to line DR, say at Q. For the bevel for joint D, strike an arc from T touching tangent ED at U to intersect ND at V. Connect VQ for the required bevel at V for joint D. For the bevel for joint H, strike an arc from T touching parallel DS at W to intersect ND at X. Connect XQ for the required bevel at X for joint H.

Although methods three and four are accurate for finding all bevels, including elliptical plans where one tangent is shorter than the other, I recommend that the learner adhere to the practice of finding the bevels for all plans using the primary method one principle.

The following 15 figures, in **Plates 65** and **66,** show all of the plan types and tangent pitch combinations possible in tangent handrailing, and the method of finding the bevel or bevels required to be applied to the butt joints of the handrail block. In **Figures 1** and **2,** showing level tangents in a quarter circle plan, the single bevel shown is applied to the level joint only. In all other figures, the initial step is to square the low point of the lower tangent to the

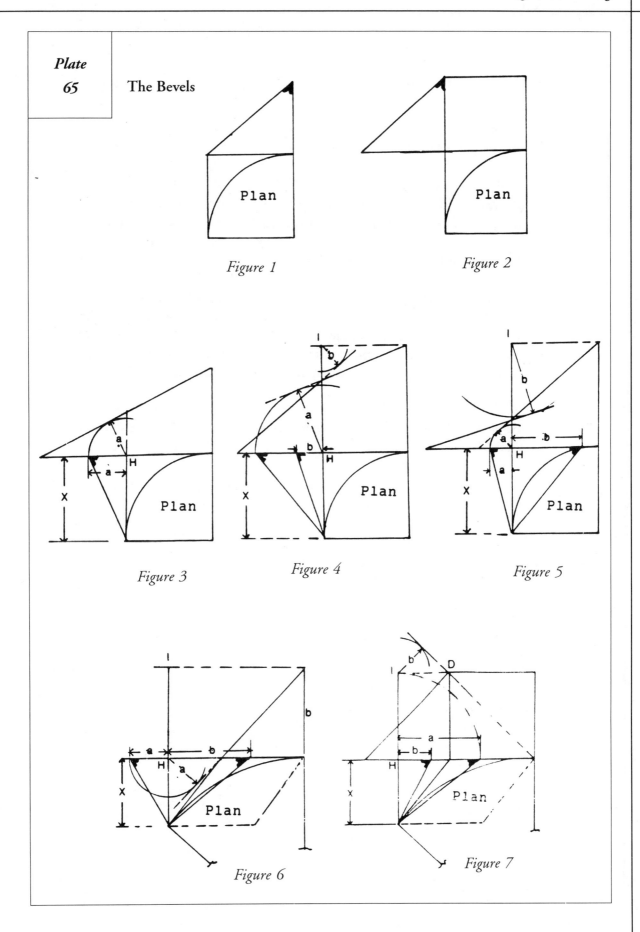

**Plate
65**   The Bevels

Figure 1

Figure 2

Figure 3

Figure 4

Figure 5

Figure 6

Figure 7

stretch out line of the plan tangents as at H, making X one leg of the right triangle to find the bevel.

To find the bevel for the upper tangent joint, from H in **Figures 3** through **15**, strike an arc touching the pitch of the upper tangent as A. A will then be the adjacent leg of the required right triangle for the bevel for the upper tangent joint. In **Figures 3, 8,** and **13**, since both tangents are the same pitch, the same bevel is required at both joints.

To find the bevel for the lower tangent joint in **Figures 6** and **11**, the height **b** will be the adjacent leg of the right triangle for the required bevel. To find the bevel for the lower tangent joint in **Figures 4, 5, 7, 9, 10, 12, 14,** and **15**, extend H to equal the height at I. Then strike an arc touching the pitch of the lower tangent as at **b** for the required leg of the triangle.

There are two instances in an obtuse plan when the tangent pitches will dictate that only one of the joints is to receive a bevel. This will be shown in **Plate 67**.

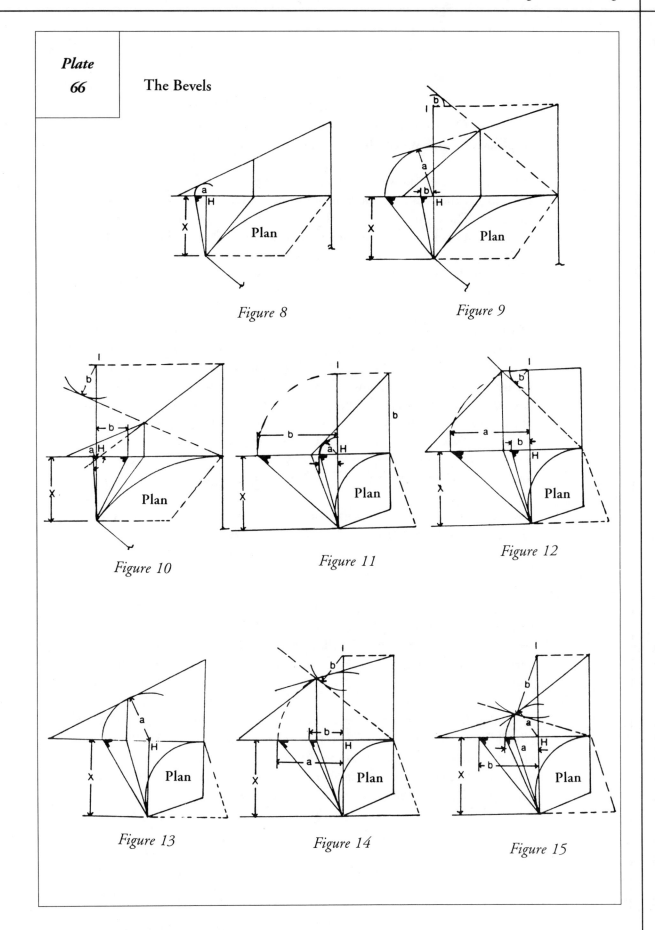

Plate 66

The Bevels

Figure 8

Figure 9

Figure 10

Figure 11

Figure 12

Figure 13

Figure 14

Figure 15

Plate 67—Obtuse Plans Showing How Tangent Pitches Affect Bevel Applications, and When One Joint Does Not Require a Bevel

It has already been shown how the pitch of the upper tangent affects the direction of the ordinate, which is not to be confused with the ordinate direction. The ordinate is always struck from the radius point at the plan. In an elliptical plan the ordinate direction, as found in **Plates 60** through **62**, is all that is required since there can be no major and minor axis from which the face mold can be drawn by using either the trammel or the string-and-pin methods.

In quarter-circle and acute plans, the ordinate always lies at, or within, the spring lines of the plan. In obtuse plans, the ordinate may lie outside of the plan spring lines, depending upon the pitch of the upper tangent. In these figures, whenever the ordinate OY lies outside of the plan spring lines, as in **Figures 1, 2, 4,** and **8,** the bevel application to the butt joint nearest to which the ordinate is pointed will be the reverse of the bevel application shown in **Figure 3, Plate 69.** This reverse bevel application is also determined whenever the pitch of the upper tangent falls below point G, or whenever the pitch of the lower tangent is above point H, also shown in **Figures 1, 2, 4,** and **8.**

Figures 3 and **7** show the ordinate to lie within the plan spring lines. **Figures 5** and **6** show the ordinate to be exactly at joints A and C respectively. In **Figure 5,** the pitch EC is shown to strike at exactly the same height as CD, at GH.

Consequently, no bevel is required for the lower butt joint F. Conversely, in **Figure 6,** the upper tangent strikes exactly at G. Therefore, no bevel is required for the butt joint at D. **Plates 99** and **100** are examples of these desirable conditions when making goosenecks or overeasements within the turn section.

In all figures, FC is the horizontal base line and GH equals height CD. Also, in all figures except **Figures 5** and **6,** two bevels are required; a different bevel for each joint. Their application to the butt joints are shown in **Plate 69,** ordinate, ordinate direction lines, minor axis, and minor axis direction lines are always shown with a small circle through the line.

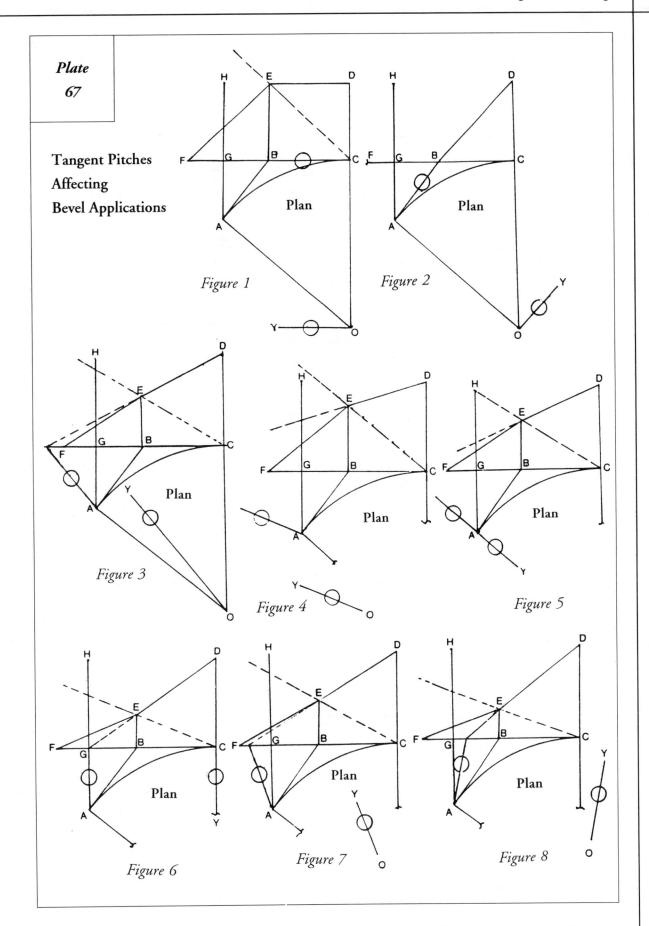

Plate 67

Tangent Pitches Affecting Bevel Applications

Figure 1

Figure 2

Figure 3

Figure 4

Figure 5

Figure 6

Figure 7

Figure 8

Plate 68—Laying Out and Squaring an Incline-Turn Section

Following is an example of the number of steps required to make an incline-turn handrail section through the tangent principle. **Plate 68** shows a quarter-turn plan where the pitched tangents are equal length.

Stage 1—Figure 1 shows the plan and elevation of the tangents of a quarter-turn handrail section. Plan tangents AB and BC are inclined as FE and ED, both equally pitched. CD is the height to be gained through the turn.

Stage 2—The angle of the elevated tangents is found as well as the face mold. The bevel for both joints is also found. **Figure 2** shows the bevel with the rectangular size of the rail drawn to determine half the joint widths of the face mold as JK. The rectangular size of the rail and the pitch of the bevel determine the block thickness to be H and the block width to be I. See **Plate 96**. Extra wood at each side of the face mold should always be added to allow more surface on which to draw the curves of the rail sides shown in **Figure 4**.

Stage 3—The block for the rail section is cut to the exact joints of the face mold. Joints are made square to the block surfaces and to the tangent lines marked at both surfaces. The tangents are squared through the block thickness at the butt joints as shown. The bevels are then applied through the block center, and the rectangular shape of the rail is drawn.

Stage 4—This is known as "sliding the mold." It is explained at **Plate 70**. It shows the excess wood to be removed to form the sides, or width, of the twisted rail through the turn.

Stages 5 and **6**—These stages, as well as **stage 4**, are again shown and further explained in **Plate 71**. They show the marking and dressing of all sides to form the squared handrail section now ready to be molded. **Figure 1-A** shows the squared rail as it twists throughout the turn.

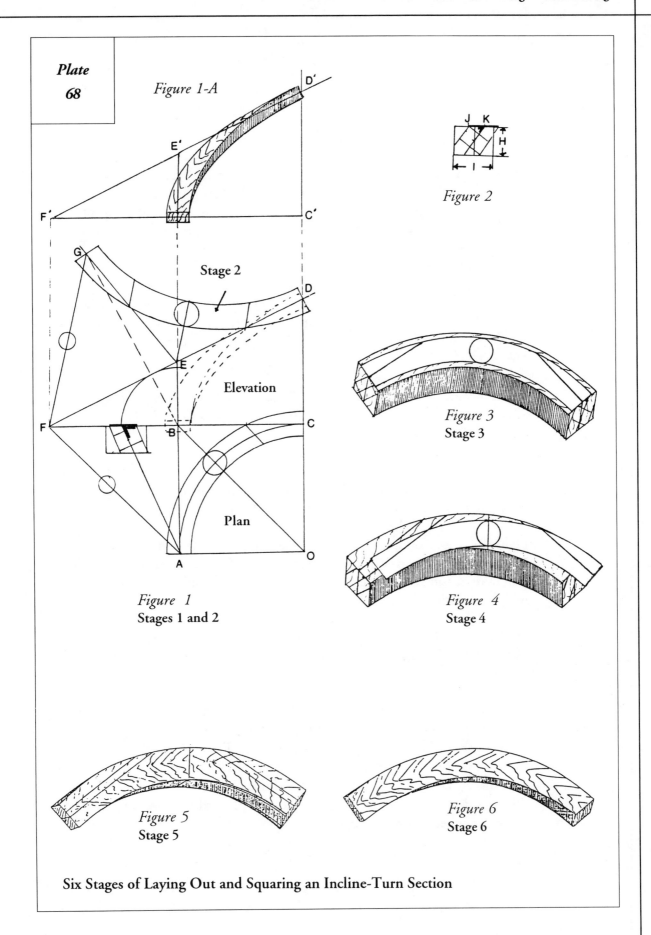

Plate 68

Figure 1-A

Stage 2

Elevation

Plan

Figure 1
Stages 1 and 2

Figure 2

Figure 3
Stage 3

Figure 4
Stage 4

Figure 5
Stage 5

Figure 6
Stage 6

Six Stages of Laying Out and Squaring an Incline-Turn Section

Plate 69—Application of the Bevels to Rail Block Joint

The bevels are always applied in a crossing manner whenever both tangents are equally pitched, or in most plans where the pitched tangents are not equal, regardless of the angle of the plan tangents. Some obtuse plans of unequally pitched tangents may require the bevels to be applied in a non-crossing manner. **Figures 1, 2, 4,** and **8** of **Plate 67** are examples. The direction of the ordinate from the radius point in these figures is shown to point outside of the plan spring lines. All other figures of **Plate 67** show the direction of the ordinate to be at or within the spring lines, as shown by the ordinate direction line. The following five figures show how the bevels are applied to the butt joints of obtuse plan handrail sections of **Plate 67**. The sections incline from left to right.

Figure 1—Refer to **Figures 1** and 4 of **Plate 67**. The dotted lower tangent pitch CE is <u>above</u> point H. Upper tangent DE extends <u>above</u> G. Ordinate direction OY is <u>outside</u> of the plan.

Figure 2—Refer to **Figures 2** and 8 of **Plate 67**. Lower tangent FB in **Figure 2** and CE in **Figure 8** extend <u>below</u> H. Upper tangent DB in **Figure 2** and DE of **Figure 8** extend <u>below</u> G. OY is <u>outside</u> of the plan.

Figure 3—Refer to **Figures 3** and 7 of **Plate 67**. Lower tangent pitch extends <u>below</u> H. Upper tangent pitch DE falls <u>above</u> G. (This is normal bevel application in right and acute angle plans regardless of tangent pitches.) OY lies <u>within</u> the plan.

Figure 4—Refer to **Figure 5** of **Plate 67**. Lower tangent pitch CE extends exactly at H. Upper tangent pitch extends <u>above</u> G. Ordinate AY strikes exactly at A, therefore no bevel will be required at joint F.

Figure 5—Refer to **Figure 6** of **Plate 67**. Upper tangent pitch strikes exactly at G. Lower tangent pitch CE extends below H. Ordinate CY strikes exactly at C, therefore no bevel will be required at joint D. The rule to follow for application of the bevel to the butt joints is: If the ordinate direction lies within the plan, the bevels will cross, as in **Figure 3** of this plate. If the ordinate direction falls outside of the plan, then the joint closer to the ordinate direction will have its bevel applied in a reverse manner.

Application of the Bevels to Rail Block Joint

Figure 1

See Figures 1 and 4 of Plate 67.
Dotted lower tangent pitch CE is <u>above</u> point H. Upper tangent DE extends <u>above</u> G. Or the direction of the ordinate OY is outside the plan.

Figure 2

See Figures 2 and 8 of Plate 67.
Lower tangent FB in Figure 2 and CE in Figure 8 extend <u>below</u> H. Upper tangent DB in Figure 2 and DE in Figure 8 are <u>below</u> G. OY is <u>outside</u> the plan.

Figure 3

See Figures 3 and 7 of Plate 67.
Lower tangent pitch CE extends <u>below</u> H. Upper tangent DE extends <u>above</u> G. (This is normal application in right and acute angle plans.) OY lies within the plan.

Figure 4

See Figure 5 of Plate 67.
Low tangent pitch extends <u>exactly</u> at H. Upper tangent pitch extends <u>above</u> G. Ordinate AY strikes at A. No bevel at joint F.

Figure 5

See Figure 6 of Plate 67..
Upper tangent pitch extends exactly at G. Lower tangent pitch CE falls below H. Ordinate CY strikes at C. No bevel at joint D.

Plate 70

Sliding the Face Mold on the Rail Block

Once the face mold is obtained and the rail block is gotten out, the bevels are applied and the new tangent lines marked at both sides. The next step is to apply the face mold along the new tangent lines marked by the bevels in order to determine the material to be removed to form the side of the incline-turn section.

Assume the handrail block shown in this plate is that of a quarter-turn section inclining from left to right. Tangents 2 and 4 of the face mold are marked on the block surfaces. The bevel shifts tangents 2 and 4 to 2' and 4' at both sides as new tangent lines. The face mold is then placed on both surfaces and shifted so that its tangents, 2 and 4, align with the new tangent lines 2' and 4' as shown. This is referred to as "sliding the face mold." The face mold is then marked along its sides to indicate the excess stock to be removed in order to form the proper twist as the rail block stands over the plan. The position of the minor axis is marked at both sides as OO, indicating the plumbline direction in which the material is removed to form the sides. Once dressed, the center of OO is marked as the center of the rail width through the turn. In some obtuse sections, where the minor axis lies beyond the spring lines, OO cannot be marked. Whenever this occurs, any convenient mark on the face mold transferred to the block serves the same purpose of determining the direction in which to dress the sides.

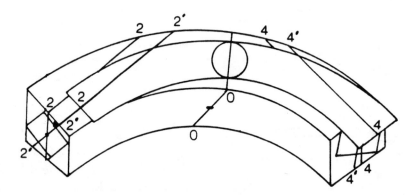

Isometric view of the sliding of the face mold

<table>
<tr><td>

Plate

71

</td><td>

Three Stages of Making the Rail Section From a Solid Block

</td></tr>
</table>

This plate is an example of forming the twist of an incline-turn handrail section from left to right from the rail block cut to the length of the face mold much like that shown in **Plate 68,** except that straight wood is extended beyond the spring lines to the joints. Once the face mold is obtained for this section, the rail block is roughly cut wider than the face mold width, say 3/4" wider at each side. The block joints are cut exactly to the joints of the face mold and squared to both block surface and the marked tangent lines.

First stage, **Figure 1**—With the joints dressed and the bevels marked through the center of the block at the butt joints, the new tangent lines are marked and the face mold applied in the sliding manner of **Plate 70** to mark the excess wood to be removed to form the sides.

Second stage, **Figure 2**—Once the sides are dressed, the transition easement of the rail thickness is drawn from joint-to-joint and to any adjoining section.

Third stage, **Figure 3**—This final stage before molding shows the rail "squared", with all four sides dressed. The rail profile can now be made.

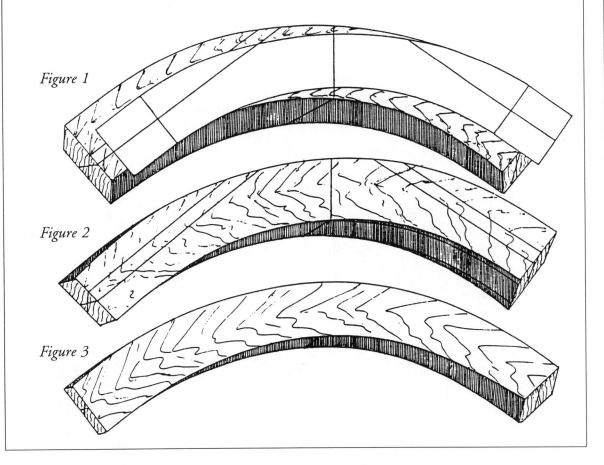

Figure 1

Figure 2

Figure 3

Plates 72 through 88—Reference Chart Showing the Layouts for Finding the Face Mold and Bevels for All Plan and Tangent Pitch Combinations in Tangent Handrailing Using Two Different Methods, "A" and "B"

The preceding plates of making incline-turn handrail sections have shown right, obtuse, and acute angle plans formed by tangents, the stretch out of the tangents along the horizontal base line, and the pitch, or pitches, of tangents. They display the ordinate and ordinate direction lines for drawing the face mold. They also show the method of obtaining the bevels and the method of sliding the face mold in order to mark the excess stock to be removed to form the twist of a handrail section.

Plate 72 shows the reference chart for the sixteen "working plates." Because the plates indexed by this chart will likely be referred to often, at each plate I have included the instructions and reference plates for each step taken.

Each of the plates will show the two methods, "A" and "B", for finding the face mold, the angle of the tangents, and the bevels. Short parallel lines to the elevated tangent's spring lines are shown should it be desired to find the face mold width along the parallels, as in **Plate 58**. Otherwise, the joint widths are found at the bevels, as in **Plate 68**. The face molds at both methods "A" and "B" represent the bottom of the rail block. The bevels are applied through the rail center at the butt joints from either surface of the block.

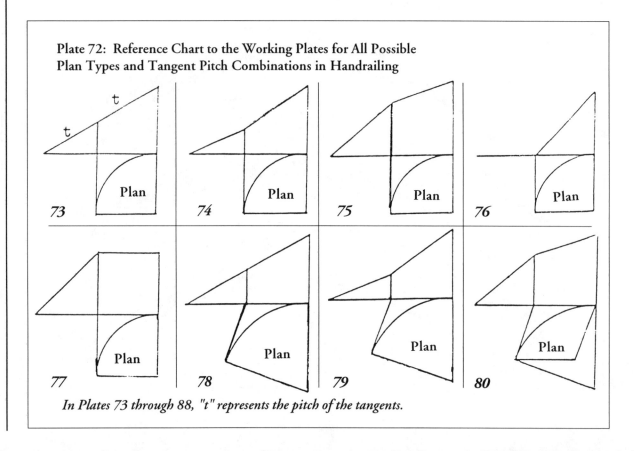

Plate 72: Reference Chart to the Working Plates for All Possible
Plan Types and Tangent Pitch Combinations in Handrailing

In Plates 73 through 88, "t" represents the pitch of the tangents.

Once the tangent pitches have been determined by the stretch out and elevation of the risers from their respective intersection points at the plan tangents, it becomes a simple matter to refer to this chart to quickly find the plan type and tangent pitch combination pertaining to the problem at hand.

Common to all of the layouts are plan tangents AB and BC, the unfolding and stretching out of the plan tangents as the horizontal base line, the ordinate and minor axis direction lines, the ordinate and minor axis, and the random parallels at the plan being transferred to the oblique plane in method "A", and at the pitch at method "B", in order to find the face mold curves at both methods. Remember, minor axis parallels will lie directly over their respective ordinate parallels at the plan, both being the same length and level.

Since only one of the two methods, either "A" or "B", is actually used, the student should not be confused by the numerous lines shown to draw both methods. Each method has its merits, but I have chosen to adhere to the method "A" principle whenever possible simply because it can be drawn within a compact area using the already established upper pitched tangent.

It is critical that the angle of the tangents, the bevels, and the ordinate and minor axis direction lines are proven to be correct before proceeding with the layout. If either one of the four is found to be done incorrectly, and is discovered once the rail block is gotten out, the block, in all likelihood, will have to be discarded.

Plate 72: Reference Chart to the Working Plates

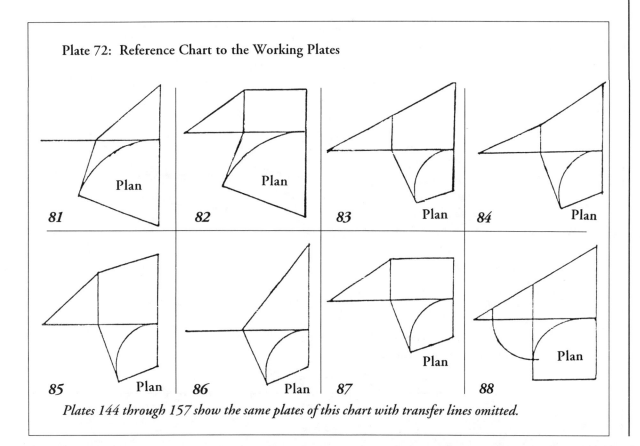

81   82   83   84

85   86   87   88

Plates 144 through 157 show the same plates of this chart with transfer lines omitted.

Plate
73   Quarter-Circle Plan with Equally Pitched Tangents

Method "A"
1. Ordinate and minor axis directions: FA and FG are equal. **Plate 60**
2. Angle of tangents: GED. BG is perpendicular to ED. EG equals EF. **Plate 53**
3 The bevel: HB equals arc touching DE. H is the bevel ¼ for both joints. **Plate 65**
4. Face mold widths along parallels DD' and GG'. **Plate 48**
5. Face mold joint widths: Found along FB at bevel. **Plate 68**
6. Face mold curves: Transfer parallels from plan. **Plate 58**
7. Block thickness and width: J and I at bevel. **Plate 68**

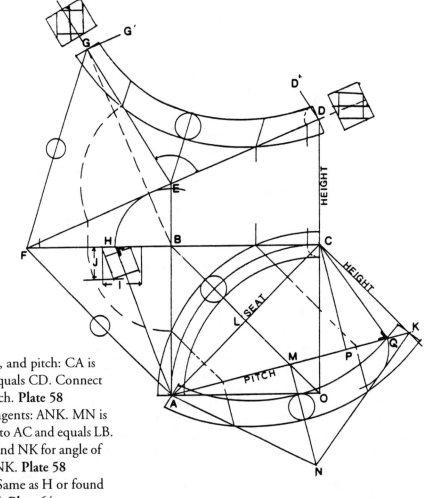

Method "B"
1. Seat, height, and pitch: CA is
the seat, CK equals CD. Connect
AK for the pitch. **Plate 58**
2. Angle of tangents: ANK. MN is
perpendicular to AC and equals LB.
Connect NA and NK for angle of
¼ tangents ANK. **Plate 58**
3. The bevel: Same as H or found
as Q along AK. **Plate 64**
4. Face mold joint widths: Found along FB at bevel H or by plan parallels from
AK. **Plate 68**
5. Face mold curves: Transfer parallels from plan. **Plate 58**
6. Block thickness and width: J and I at bevel H. **Plate 68**

Plate 74

Quarter-Circle Plan with Short Lower Pitched Tangent

Method "A"
1. Ordinate and minor axis direction: HA and HG are equal. **Plate 60**
2. Angle of tangents: GED. BG is perpendicular to ED, EG equals EF. **Plate 55**
3. The bevels: BJ equals CD. BI equals arc from B touching tangent pitch DE for the bevel for joint D. BK equals arc from J touching tangent pitch FE for the bevel for joint G. **Plate 65**
4. Face mold widths along parallels DD' and GG': **Plate 50**
5. Face mold joint widths: Found along FB at bevels K and I. **Plate 68**
6. Face mold curves: Transfer parallels from plan. **Plate 58**
7. Block thickness and width: M and L at bevels. **Plate 68**

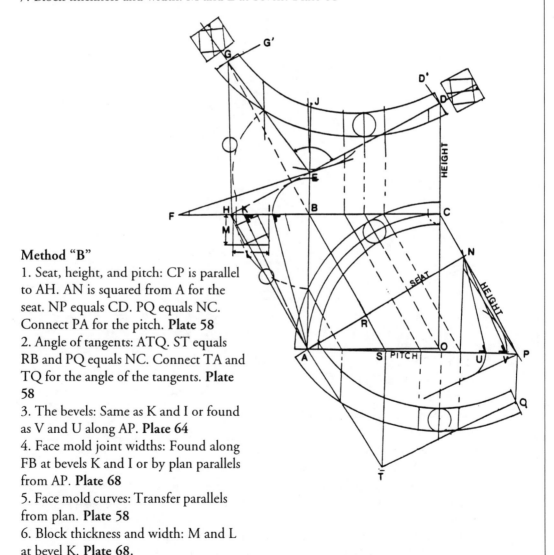

Method "B"
1. Seat, height, and pitch: CP is parallel to AH. AN is squared from A for the seat. NP equals CD. PQ equals NC. Connect PA for the pitch. **Plate 58**
2. Angle of tangents: ATQ. ST equals RB and PQ equals NC. Connect TA and TQ for the angle of the tangents. **Plate 58**
3. The bevels: Same as K and I or found as V and U along AP. **Plate 64**
4. Face mold joint widths: Found along FB at bevels K and I or by plan parallels from AP. **Plate 68**
5. Face mold curves: Transfer parallels from plan. **Plate 58**
6. Block thickness and width: M and L at bevel K. **Plate 68.**

Plate 75

Quarter-Circle Plan with Short Upper Pitched Tangent

Method "A"
1. Ordinate and minor axis directions: HA and HG are equal. **Plate 60**
First alternate method: BI equals CD. OK equals IJ. Connect KA. **Plate 63**
2. Angle of tangents: GED. BC is perpendicular to DE, EG equals EF. **Plate 54**
3. The bevels: BI equals CD. BL equals arc from I touching pitch FE for the bevel for joint G. BM equals arc from B touching pitch DE for the bevel for joint D **Plate 65**
4. Face mold widths along parallels DD' and GG': **Plate 49**
5. Face mold joint widths: Found along FB at bevels M and L. **Plate 68**
6. Face mold curves: Transfer parallels from plan. **Plate 58**
7. Block thickness and width: 1 and 2 at bevel M. **Plate 68**

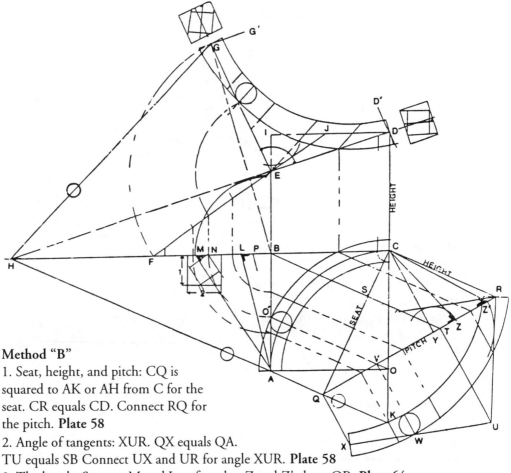

Method "B"
1. Seat, height, and pitch: CQ is squared to AK or AH from C for the seat. CR equals CD. Connect RQ for the pitch. **Plate 58**
2. Angle of tangents: XUR. QX equals QA. TU equals SB Connect UX and UR for angle XUR. **Plate 58**
3. The bevels: Same as M and L or found as Z and Z' along QR. **Plate 64**
4. Face mold joint widths: Found along HB at bevels M and L or by plan parallels from QR. **Plate 68**
5. Face mold curves: Transfer parallels from plan. **Plate 58**
6. Block thickness and width: 1 and 2 at bevel M. **Plate 68**

Plate 76

Quarter-Circle Stair—The Upper Tangent Is Pitched, the Lower Tangent Is Level

Method "A"
1. Ordinate and minor axis directions: BA and BE are equal. **Plate 60**
2. Angle of tangents: EBD. BE equals BA. **Plate 57**
3. The bevels: B or D. **Plate 65**
4. Face mold joint width at E: Draw parallels from BD at bevel. **Plate 68**
5. Face mold curves: Transfer parallels from plan. **Plate 58**
6. Block thickness and width: I and J at bevel B. **Plate 68**

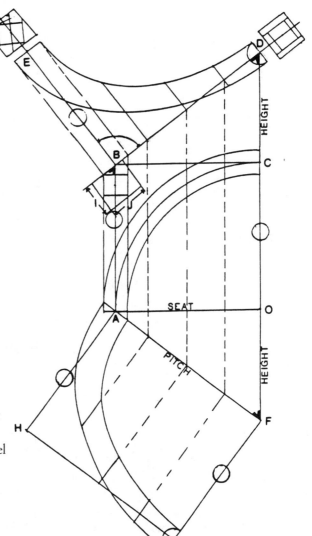

Method "B"
1. Seat, height, and pitch: AO is seat, FO is height, and FA is the pitch. **Plate 58**
2. Angle of tangents: AH and FG are square to AF. AH equals AB and FG equals AH. Connect GH for angle GHA. **Plate 58**
3. The bevels: Same as at B or D or at F along AF. **Plate 65**
4. Face mold joint width at A: At bevel B or by plan parallels from FA. **Plate 68**
5. Face mold curves: Transfer parallels from plan. **Plate 58**
6. Block thickness and width: I and J at bevel B. **Plate 68**

Plate 77

Quarter-Circle Stair—The Lower Tangent Is Pitched, the Upper Tangent Is Level

Method "A"

1. Ordinate and minor axis directions: BC and ED are equal. **Plate 60**
2. Angle of tangents: DEG. EG equals EF, DO' equals EG. **Plate 56**
3. The bevels: Bevel at E. **Plate 68**
4. Face mold joint width at D: Bevel E. **Plate 68**
5. Face mold curves: Transfer parallels from plan or use the trammel method. By trammel: For inside curve, 5-6 equals 2-0', and 5-7 equals 3-0'. For outside curve, 8-9 equals 1-0', and 8-10 equals 4-0'. **Plate 58**
6. Block thickness and width: Y and Z at bevel E. **Plate 68**

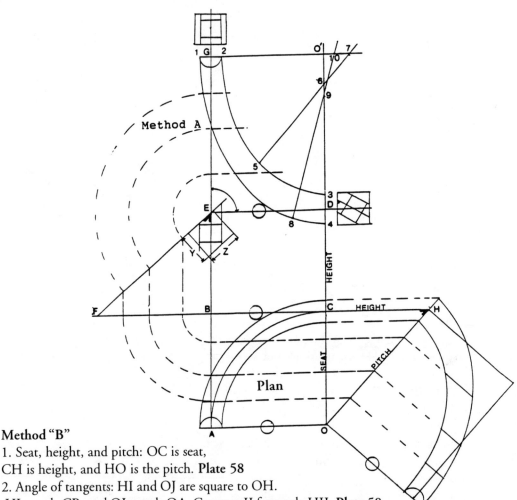

Method "B"

1. Seat, height, and pitch: OC is seat,
CH is height, and HO is the pitch. **Plate 58**
2. Angle of tangents: HI and OJ are square to OH.
HI equals CB, and OJ equals OA. Connect IJ for angle HIJ. **Plate 58**
3. The bevels: Same as at E or along OH. **Plate 64**
4. Face mold joint width at H: Found at bevel E or from plan parallels along OH. **Plate 68**
5. Face mold curves: Transfer parallels from plan. **Plate 58**
6. Block thickness and width: Y and Z at bevel E. **Plate 68**

Obtuse Plan—Equally Pitched Tangents

Method "A"
1. Ordinate and minor axis directions: FA and FG are equal. **Plate 61**
2. Angle of tangents: GED. HG is perpendicular to ED, EG equals EF. **Plate 53**
3. The bevel.: HI equals arc touching FE at I. **Plate 66**
4. Face mold widths at parallel DD'. **Plate 48**
5. Face mold joint widths: Found along FB at bevel I. **Plate 68**
6. Face mold curves: Transfer parallels from plan. **Plate 58**
7. Block thickness and width: Y and Z at bevel I. **Plate 68**

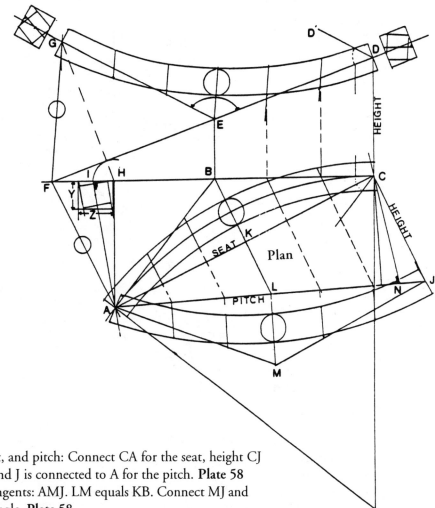

Method "B"
1. Seat, height, and pitch: Connect CA for the seat, height CJ
 equals CD, and J is connected to A for the pitch. **Plate 58**
2. Angle of tangents: AMJ. LM equals KB. Connect MJ and
 MA for the angle. **Plate 58**
3. The bevel: Same as I or found along AJ as N. **Plate 64**
4. Face mold joint widths: Found along FB at bevel I or by plan parallels from AC.
Plate 68
5. Face mold curves: Transfer parallels from plan. **Plate 58**
6. Block thickness and width: Y and Z at bevel. **Plate 68**

Plate 79

Obtuse Plan—Short Lower-Pitched Tangent

Method "A"

1. Ordinate and minor axis directions: JA and JG are equal. **Plate 61**
2. Angle of tangents: GED. HG is perpendicular to ED, EG equals EF. **Plate 55**
3. The bevels: HI equals height CD, CE equals pitch FE, arc from I touching CE is HT for the bevel for joint G, arc from H touching extended DE is HS for the bevel for joint G. **Plate 66**
4. Face mold widths along parallels DD' and GG'. **Plate 50**
5. Face mold joint widths: Found along FB at bevels T and S. **Plate 68**
6, Face mold curves: Transfer parallels from plan. **Plate 51**
7. Block thickness and width: Y and Z at bevel T. **Plate 68**

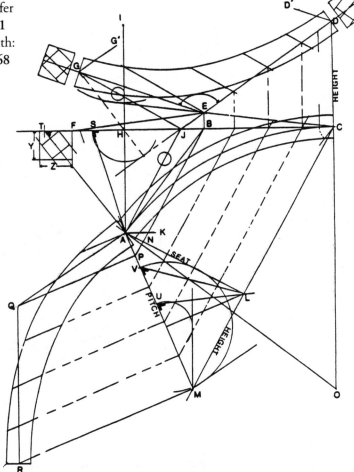

Method "B"

1. Seat, height, and pitch: CL is parallel to AJ, AL is square to AJ for the seat, height LM equals CD, and MA is connected for the pitch. **Plate 58**
2. Angle of tangents: AQR. MR equals LC and PQ equals NB.Connect QA and QR for the angle AQR. **Plate 58**
3. The bevels: Same as at T and S or found as U and V along AM. **Plate 64**
4. Face mold joint widths: Found along FB at bevels S and T or by plan parallels from AM. **Plate 68**
5. Face mold curves: Transfer parallels from plan. **Plate 58**
6. Block thickness and width: Y and Z at bevel T **Plate 68**

Plate 80

Obtuse Plan—Short Upper-Pitched Tangent

Method "A"

1. Angle of tangents: GED. HG is perpendicular to DE, EG equals EF. **Plate 54**
2. Ordinate and minor axis directions (first alternate method): Draw plan parallelogram ABCN. NP equals LM. PA is ordinate direction. NQ is parallel to PA. GEDS is the oblique plane parallelogram. RS is the minor axis direction. **Plate 63**
3. The bevels: HI equals CD, CE equals pitch FE, arc from I touching pitch CE is HJ for the bevel for joint G. Arc from H touching pitch DE is HK for the bevel for joint D. **Plate 66**
4. Face mold widths along parallels DS and GS. **Plate 49**
5. Face mold joint widths: Found along FB at bevels F and J. **Plate 68**
6. Face mold curves: Transfer parallels from plan. **Plate 58**
7. Block thickness and width: Y and Z at bevel K. **Plate 68**

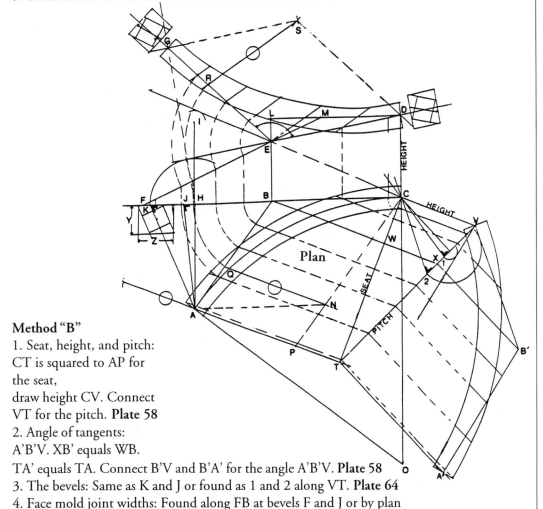

Plan

Method "B"

1. Seat, height, and pitch: CT is squared to AP for the seat, draw height CV. Connect VT for the pitch. **Plate 58**
2. Angle of tangents: A'B'V. XB' equals WB. TA' equals TA. Connect B'V and B'A' for the angle A'B'V. **Plate 58**
3. The bevels: Same as K and J or found as 1 and 2 along VT. **Plate 64**
4. Face mold joint widths: Found along FB at bevels F and J or by plan parallels from VT. **Plate 68**
5. Face mold curves: Transfer parallels from the plan. **Plate 58**
6. Block thickness and width: Y and Z at bevel K. **Plate 68**

Plate 81	Obtuse Plan—The Lower Tangent Is Level, the Upper Tangent Is Pitched

Method "A"

1. Ordinate and minor axis directions: BA and BF are equal. **Plate 60**
2. Angle of tangents: FBD. FE is perpendicular to BD. **Plate 57**
3. The bevels: G and L. EL equals CD for the bevel for joint F. Arc from E touching pitch DB at G is the bevel for joint D. **Plate 65**
4. Face mold widths along parallels DD' and FF'. **Plate 52**
5. Face mold joint widths: Found along CB at bevels L and G. **Plate 68**
6. Face mold curves: Transfer parallels from plan. **Plate 58**
7. Block thickness and width: Y and Z at bevel L. **Plate 68**

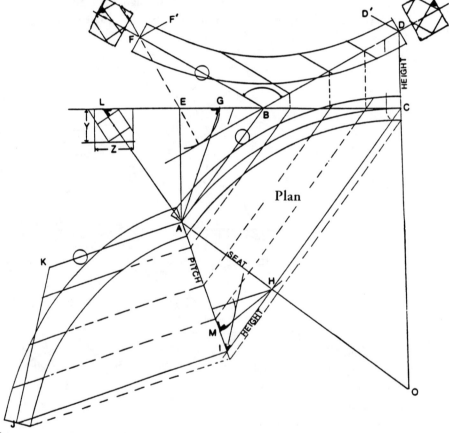

Plan

Method "B"

1. Seat, height, and pitch: CH is parallel to AB. HI is the height, IA is the pitch. **Plate 58**
2. Angle of tangents: AKJ. AK equals AB and IJ equals HC. Connect KJ and KA for the angle. **Plate 58**
3. The bevels: Same as G and L or found as M and I along AI. **Plate 64**
4. Face mold joint widths: Found along CB at bevels G and L or by plan parallels from AI. **Plate 68**
5. Face mold curves: Transfer parallels from plan. **Plate 58**
6. Block thickness and width: Y and Z at bevel L. **Plate 68**

Plate
82

Obtuse Plan—The Upper Tangent Is Level, the Lower Tangent Is Pitched

Method "A"
1. Ordinate and minor axis directions: BC and DE. **Plate 60**
2. Angle of tangents: GED. EG equals EF. **Plate 56**
3. The bevels: J and K. HJ equals HI for the bevel for joint D, arc from I touching pitch FE is HK for the bevel for joint G. **Plate 66**
4. Face mold widths along parallels DD' and GG'. **Plate 51**
5. Face mold joint widths: Found along BF at bevels J and K. **Plate 68**
6. Face mold curves: Transfer parallels from plan. **Plate 58**
7. Block thickness and width: Y and Z at bevel J. **Plate 68**

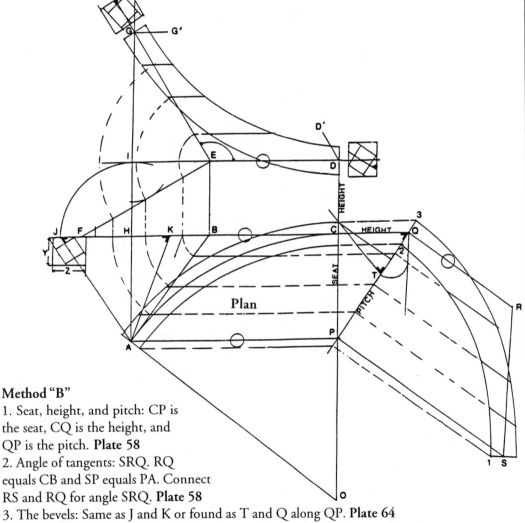

Plan

Method "B"
1. Seat, height, and pitch: CP is the seat, CQ is the height, and QP is the pitch. **Plate 58**
2. Angle of tangents: SRQ. RQ equals CB and SP equals PA. Connect RS and RQ for angle SRQ. **Plate 58**
3. The bevels: Same as J and K or found as T and Q along QP. **Plate 64**
4. Face mold joint widths: At bevels J and K or from plan parallels at QP. **Plate 68**
5. Face mold curves: Transfer parallels from plan. **Plate 58**
6. Block thickness and width: Y and Z at bevel J. **Plate 68**

Plate	
83	**Acute Plan—Equally Spaced Tangents**

Method "A"
1. Ordinate and minor axis directions: FA and FG are equal. **Plate 62.**
2. Angle of GED. GH is perpendicular to ED. EG equals EF. **Plate 56.**
3. The bevel: HI equals arc touching DE at I. **Plate 66.**
4. Face mold width along parallel DD'. **Plate 48.**
5. Face mold joint widths: Found along FB at bevel I. **Plate 68.**
6. Face mold curves: Transfer parallels from plan. **Plate 58.**

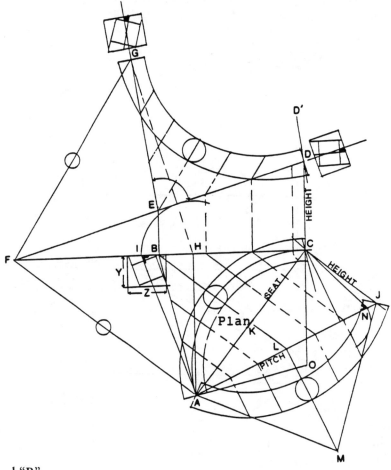

Method "B"
1. Seat, height, and pitch: Connect CA for the seat. CJ equals CD, and J is connected to A for the pitch. **Plate 58.**
2. Angle of tangents: AMJ. LM equals KB. Connect MJ and MA for the angle. **Plate 58.**
3. The bevel: Same as I or found along AJ as N. **Plate 64.**
4. Face mold joint widths: Found along FB at bevel I or by plan parallels from AC. **Plate 68.**
5. Face mold curves: transfer parallels from plan. **Plate 58.**
6. Block thickness and width: Y and Z at bevel I. **Plate 68.**

Plate 84

Acute Plan—Short Lower-Pitched Tangent

Method "A"
1. Ordinate and minor axis directions: LA and LG are equal. **Plate 62**
2. Angle of tangents: GED. GH is perpendicular to ED, EG equals EF. **Plate 55**
3. The bevels: HK equals arc from H touching DE for the bevel for joint D. HI equals CD. HJ equals arc from I touching pitch CE for the bevel for joint G. **Plate 66**
4. Face mold widths along parallels DD' and GG'. **Plate 50**
5. Face mold joint widths: Found along AB at bevels J and K. **Plate 68**
6. Face mold curves: Transfer parallels from plan. **Plate 58**
7. Block thickness and width: Y and Z at bevel J. **Plate 68**

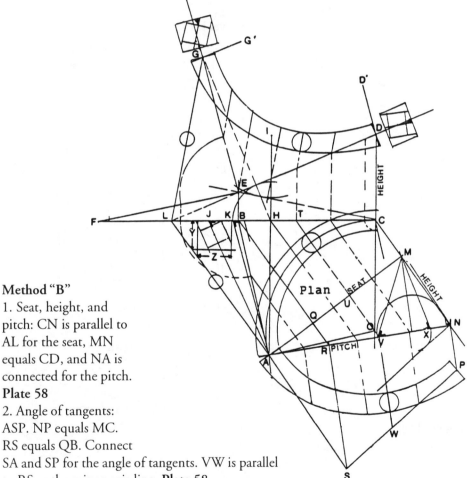

Method "B"
1. Seat, height, and pitch: CN is parallel to AL for the seat, MN equals CD, and NA is connected for the pitch. **Plate 58**
2. Angle of tangents: ASP. NP equals MC. RS equals QB. Connect SA and SP for the angle of tangents. VW is parallel to RS as the minor axis line. **Plate 58**
3. The bevels: Same as J and K or found as V and X along AN. **Plate 64**
4. Face mold joint widths: Found along FB at bevels J and K or by plan parallels from AM. **Plate 68**
5. Face mold curves: Transfer parallels from plan: **Plate 58**
6. Block thickness and width: Y and Z at bevel J. **Plate 68**

Plate

85

Acute Plan—Short Upper-Pitched Tangent

Method "A"

1. Angle of tangents: HG is perpendicular to DE. EG equals EF for angle of tangents GED. GED3 is the parallelogram. **Plate 54**
2. Ordinate and minor axis directions (first alternate method): NP equals LM. AP is ordinate direction. N1 is parallel to AP. 2-3 is minor axis direction. **Plate 63**
3. The bevels: HI equals CD. HJ equals arc from H touching ED for the bevel for joint D. HK equals arc from I touching pitch CE for the bevel for joint G. **Plate 66**
4. Face mold width along D3 and G3. **Plate 49**
5. Face mold joint widths: Found at bevels J and K. **Plate 68**
6. Face mold curves: Transfer parallels from plan. **Plate 58**
7. Block width and thickness: Y and Z at bevel J. **Plate 68**

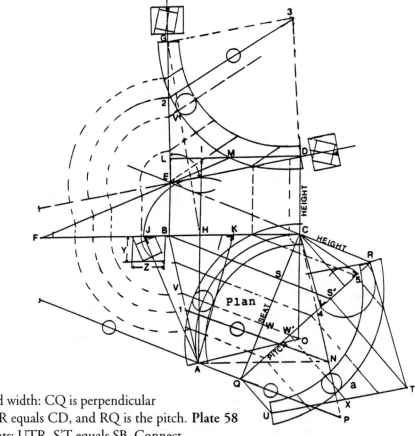

Method "B"

1 Seat, height, and width: CQ is perpendicular to AP from C, CR equals CD, and RQ is the pitch. **Plate 58**
2. Angle of tangents: UTR. S'T equals SB. Connect TU and TR for the angle of the tangents. **Plate 58**
3. The bevels: Same as J and K or found as 4 and 5 along QR. **Plate 64**
4. Face mold joint widths: At bevels F and J or by plan parallels from QR. **Plate 68**
5. Face mold curves: Transfer parallels from plan. **Plate 58**
6. Block thickness and width: Y and Z at bevel J. **Plate 68**

Plate 86

Acute Plan—The Upper Tangent Is Pitched, the Lower Tangent Is Level

Method "A"

1. Ordinate and minor axis directions: BA and BE are equal. **Plate 60**
2. Angle of tangents: EBD. EF is perpendicular to DB, and BE equals BA. **Plate 57**
3. The bevels: Arc from F touching BD is bevel G for joint D. **Plate 66**
4. Face mold widths along parallels DD' and EE'. **Plate 52**
5. Face mold joint widths: Found along FC at bevels. **Plate 68**
6. Face mold curves: Transfer parallels from plan. **Plate 58**
7. Block thickness and width: Y and Z at bevel I. **Plate 68**

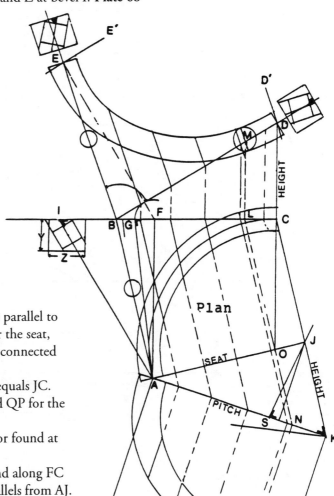

Method "B"

1. Seat, height, and pitch: CK is parallel to AB. AJ is squared to AB at A for the seat, height JK equals CD, and KA is connected for the pitch. **Plate 58**
2. Angle of tangents: AQP. KP equals JC. AQ equals AB. Connect QA and QP for the angle AQP. **Plate 58**
3. The bevels: Same as I and G or found at AK as S and K. **Plate 64**
4. Face mold joint widths: Found along FC at bevels I and G or by plan parallels from AJ. **Plate 68**
5. Face mold curves: Transfer parallels from plan. **Plate 58**
6. Block thickness and width: Y and Z at bevel I. **Plate 68**

Plate 87

Acute Plan—The Upper Tangent Is Level, the Lower Tangent Is Pitched

Method "A"
1. Ordinate and minor axis directions: BC and DE. **Plate 60**
2. Angle of tangents: GED. EG equals EF at G. **Plate 56**

3. The bevels: HK equals HI for the bevel for joint D. HJ equals arc from I to pitch CE for the bevel for joint G. **Plate 66**
4. Face mold widths along parallels DD' and GG': **Plate 51**
5. Face mold joint widths: Found along BF at bevels J and K. **Plate 68**
6. Face mold curves: Transfer parallels from plan. **Plate 58**
7. Block thickness and width: X and Y at bevel K. **Plate 68**

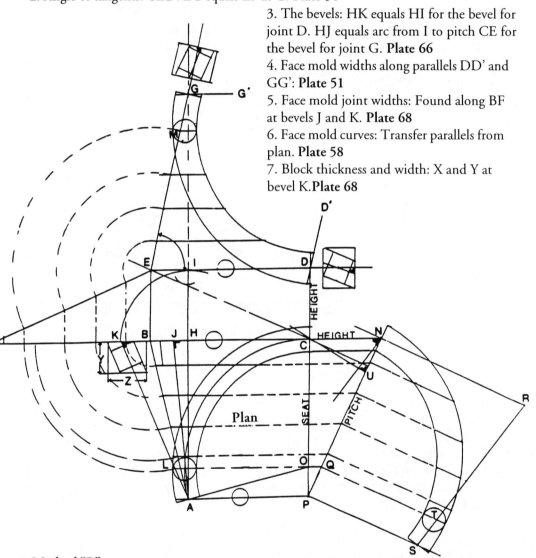

Plan

Method "B"
1. Seat, height, and pitch: CO is seat, CN is height, and NP is the pitch. **Plate 58**
2. Angle of tangents: SNR. NR equals BC. PS equals PA. Connect RN and RS for the angle SNR. **Plate 58**
3. The bevels: Same as J and K or found as N and U along PN. **Plate 64**
4. Face mold joint widths: At bevels J and K or from plan parallels along PN. **Plate 68**
5. Face mold curves: Transfer parallels from plan. **Plate 58**
6. Block thickness and width: Y and Z at bevel K. **Plate 68**

| Plate 88 | The Elliptical Plan |

In radius-drawn plans both tangents touch the centerline curve at an equal distance from the vertex. In an elliptical plan such as this, the tangents will touch the plan curve at different distances from the vertex. Consequently, both tangents will not be equal length.

Finding the angle of the elevated tangents, the bevels, and the face mold is quite simple if method "A" is used. Make both plan tangents equal length for layout purposes, then reduce the shortened tangent to its proper length at the pitch of the tangent and at the face mold. The shortened tangent will not effect any change in the bevel, since a bevel applies to the entire length of a tangent.

In this plan, let the shortened plan tangent BA be increased to equal BC as BA'. The layout is then made as though both tangents were equal length. FC is the stretch out line of the plan tangents. **Plate 74** is the layout for this plan. With the angle of the tangents found, tangent EG can then be shortened to EH as shown. The width of the face mold joints is found by either the bevels, as in **Plate 68,** or by the oblique plane parallels, as in **Plates 48** through **52.** DL is equal to FK, and HJ is equal to DI. By making both plan tangents equal length, there will be no deviation in the method of finding the bevels.

However, should the railmaker wish to use the method "A" principle of finding the bevels without increasing the length of the short plan tangent, the same bevels can be found by a more intricate method at the angle of the elevated tangents, as shown in **Plate 89.** Therefore, in using the method "A" principle, do not attempt to find the bevels along the stretch out line of the plan tangents without first increasing the length of the shortened plan tangent.

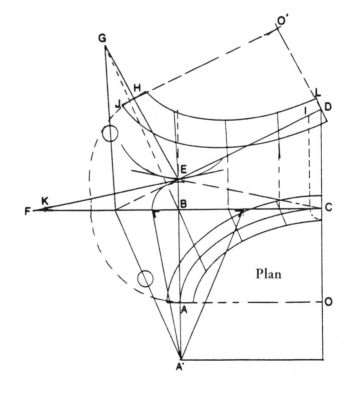

Plan

Plate

89

Finding the Bevels for an Elliptical Plan in Method "A" without Increasing the Length of the Shortened Plan Tangent (Review of Figure 5, Plate 64)

In **Plate 88** the shortened plan tangent is increased to equal the other tangent in order to find the bevels in the manner shown in **Figure 10, Plate 66.** This plate shows a similar obtuse plan where the bevels are found at the angle of the tangents in method "A" without the necessity of increasing the short plan tangent. Its reliance depends upon the accuracy of the minor axis direction.

To find the minor axis direction line, extend DE to strike the horizontal base line at F. Connect FA and extend. With C as a radius point and arc CI as the radius, strike an arc touching FC at J. With D as a radius point and DJ as the radius, strike an arc around K. From F, draw FK touching arc DJ for the minor axis direction.

Proceed to find the angle of the elevated tangents as in **Figure 10, Plate 66.** The bevels are found in the same manner as in **Figure 5, Plate 64.** From the extension of direction FL, draw right angle FKD from D. With K as a radius point and either CI or CJ as the radius, strike an arc toward M. Draw D touching the arc at M. At any point along DK draw a perpendicular line touching extended DM, say at M. The perpendicular along DK is NM. From D, draw DQ parallel to elevated tangent EL. To find the bevel for joint L, strike an arc from N touching DQ to intersect at R. Connect RM for the required bevel NRM. For the bevel for joint D, let arc NP touch tangent DE for bevel NPM.

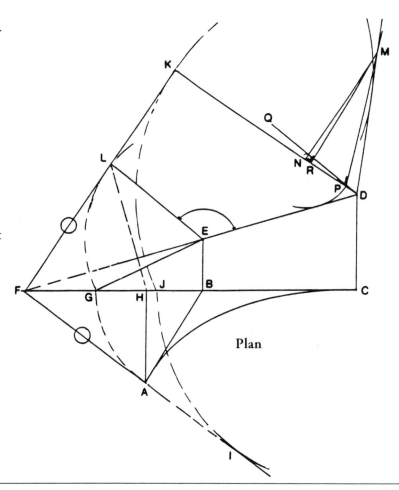

Plan

Plate 90

Finding the Bevels for an Elliptical Plan in Method "B" without Increasing the Shortened Plan Tangent (Review of Figure 4, Plate 64)

This is the same plan as in **Plate 89**. The bevels for this elliptical plan, the angle of the tangents, and the method of drawing the face mold using the method "B" principle have already been shown in **Figure 4, Plate 64** for equal length plan tangents. The method is the same when one of the plan tangents is shorter than the other. The procedure is as follows:

1. The ordinate direction: Draw FA and extend as at H.
2. The seat: Draw HC perpendicular to FH.
3. The height: Make CI equal to CD and perpendicular to HC.
4. The pitch: Connect HI as the distance from H to C at height CD.
5. The angle of the elevated tangents: Draw HJ equal to HA and perpendicular to HC. Make BKL parallel to FH. Let LM equal KB. Connect MI and MJ for the angle of the elevated tangents.
6. The bevels: From C make CN perpendicular to HI. Draw IO parallel to tangent JM. For the bevel for joint I, from N strike an arc touching MI to intersect at P. Connect PC for the bevel CPN for joint I. For the bevel for joint J, from N strike an arc touching tangent parallel IO to intersect IH at Q. Connect QC for the bevel CQN for joint J.
7. Face mold curves are found by parallels from the plan, as in **Plate 58**.

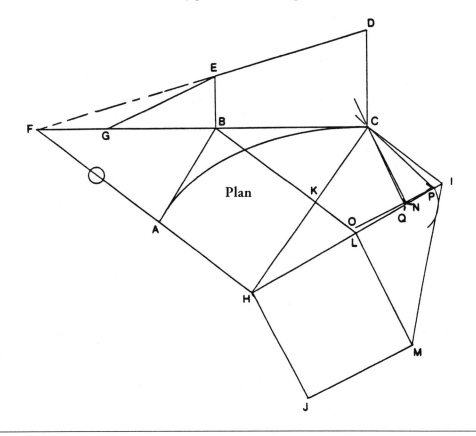

Plan

Plate 91

Alternate Method "B" of Finding the Face Mold Can Be Drawn at Any Accessible Area

The face mold, as shown in method "B" in **Plate 58**, is drawn at the plan by ordinate parallels. Whenever there is not enough room to draw the layout below the plan, it may also be drawn at a more convenient area by simply transferring the right angle formed by the seat and height.

Figure 1 is the same plan and tangent pitch combination as in **Plate 75**, but only the seat is shown. The angle of the tangents and the face mold is to be drawn elsewhere.

Figure 2 shows the right angle of the seat CM and height CD', CD' being equal to CD in **Figure 1**. With CM marked as in **Figure 1**, the angle of the elevated tangents, as well as the face mold shape, can be drawn. The joint widths of the face mold are found through parallels 1 and 4 as WX and YZ respectively. RS, UV, and MT are the same lengths as PB, NQ, and AM in **Figure 1**, as are parallels I through 4.

Whenever the face mold of a radius drawn plan is to be found at an open area, it can be drawn by either the trammel or string and pin method. With the minor axis VO' the same as QO in **Figure 1**, the major axis line can be drawn perpendicular at O'. For the inside curve using the trammel method, let Y-7 equal 5-0' and mark 8 on the same rod. For the outside curve, let X-10 equal 6-0' and mark 11 on the same rod. Proceed as in **Figure 1, Plate 58**.

To find pin placements to draw the face mold curves with the string and pin method shown in **Figure 1, Plate 58**, let 5-9 equal Y-8 for the inside curve. For the outside curve, let 6-12 equal X-11. Points 9 and 12 are then the pin placements.

Since drawing the face mold using either the trammel or string and pin method requires ample room in large-radius plans, for all practical purposes, I recommend the learner adhere to the practice of finding the face mold shape using ordinate parallels. All bevels are found using the primary method shown at the stretch out line of the plan tangents.

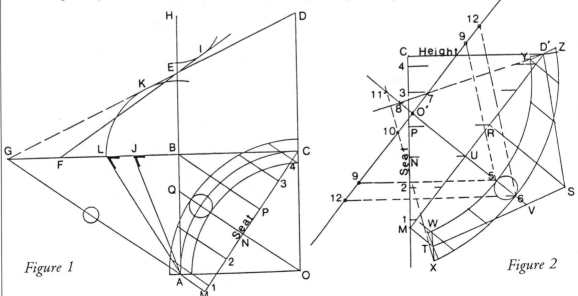

Figure 1

Figure 2

Plate
92

The Full Round Handrail Section

A full round incline-turn handrail section is the simplest of all rail sections to make. Because the rail is a full round, there is not a rectangular twisted section to be made.

The pattern for the section is not a typical face mold, but rather an equal-width elliptical curved pattern of the rail diameter throughout the curve; the block thickness is simply that of the rail diameter. The bevel, or bevels, must still be applied to the butt joints as in a molded type rail section, and the joints must be squared to both the tangent lines and the block surface.

This is an example of a plan and elevation of such a full round quarter-turn handrail section with equally pitched tangents. The rail centerline is first drawn at the pitched tangents' angle then half the diameter is set off at each side as shown.

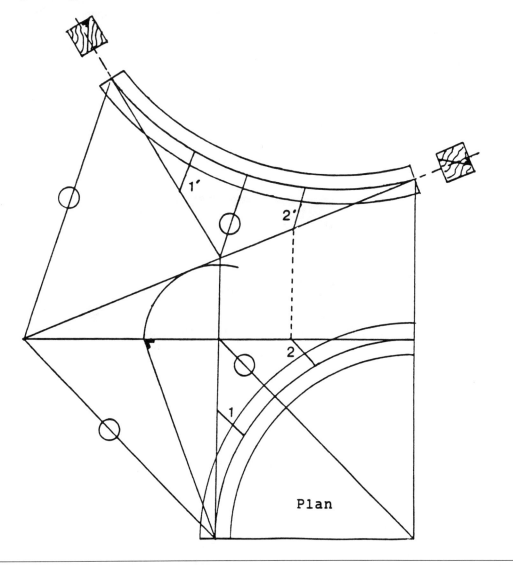

Plan

Plate	
93	Squaring and Shaping the Handrail Section

An incline-turn handrail section is first made to the rectangular size of the rail thickness and width , a process known as "squaring." This can be done efficiently by use of the bandsaw, disc and spindle sanders with coarse grit paper or cloth, and hand tools. Once squared, the rail is ready to be shaped to the desired profile. Needless to say, the simpler the profile, the faster the job is done, and the best equipment available will pay for itself in labor saved. The router, spindle carver, shaper, and small thumb planes can be used to tool the profile. Sanding blocks shaped to the desired contour are an asset in finishing the job.

Should a plow be required at the bottom of the rail to receive balusters, there must also be a fillet member to fit into the plow between the balusters. If the twist permits, laminate thin material to the thickness of the fillet at the rail bottom between the balusters prior to making the plow. Once the plow is made, the fillet can then be fitted.

This plate shows an example of a rectangular "squared" rail to be shaped to the profile shown. The shaded areas should first be removed by scribing positions A through E at both sides of the rail. The contours can then be shaped and sanded as shown.

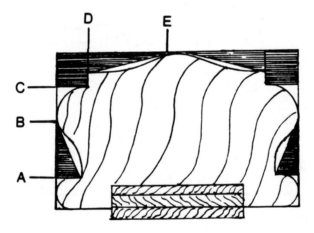

Craftsman dressing two incline-turn 4"x6" red oak handrail sections through the connecting joint prior to shaping to the desired profile. The wood grain is the outstanding feature of this tangent-made handrail.

Plate 94—The Joint, Rail Bolt, and Side View of Top and Bottom Rail Curves

Figure 1 shows a typical profile of a handrail butt joint to be joined with another section. The center hole is to receive a handrail screw (rail bolt) used to tighten the joint between two handrail sections. Two side dowels are used to further strengthen and align the two joined sections.

Figure 2 shows a typical handrail screw. It is approximately 4-½" in length and 5/16" to 3/8" in diameter. The nut for the machine screw end is notched so that it can be tightened with a punch or a common carpenter's nail-set.

Figure 3 shows two handrail sections joined together with the handrail screw. Whenever two sections of handrail are to be joined, the rail size at the joint should be slightly in excess of the net size so that a smooth transition can be worked between the two sections.

With the use of the face mold, the sides of the two sections to be joined are marked and dressed, leaving the joint between the two sections slightly wider. Likewise, at the joint where the machine screw portion is to be freely inserted, the bottom of the rail should be cut approximately ¼" below the estimated finished line of the bottom of the rail approximately 6" from the joint so as to allow a nut tightening hole to be bored at the bottom.

The lag portion of the handrail screw is threaded into a pre-bored hole at the joint of one of the two sections. Although I have shown in **Plate 94** the lag end inserted into the joint of the upper section, it is always better to have it in the lower section so as to make it easier to start the tightening nut through the bottom hole in the upper section. However, because the pitch of this rail is very low, there would be little difficulty in threading the tightening nut.

The protruding bolt length end with the machine thread is marked at the bottom side of the adjoining section. At the end point, a nut-tightening hole is bored up into the rail slightly past the rail center. At this point, a hole large enough to freely receive the bolt is bored from the butt end to the pre-bored bottom hole. The bottom hole must be bored deep enough to be able to insert and tighten the nut, while being careful not to bore too deeply so as to interfere with the profile at the top of the rail. Unless a special bored insert washer with one side curved to the diameter of the nut-tightening hole is used, the tightening hole will have to be flattened at the bolt hole so as to be able to receive the flat metal washer for the nut. The two sections can now be assembled and worked through the joint sides. The bottom line can then be marked and the rail dressed.

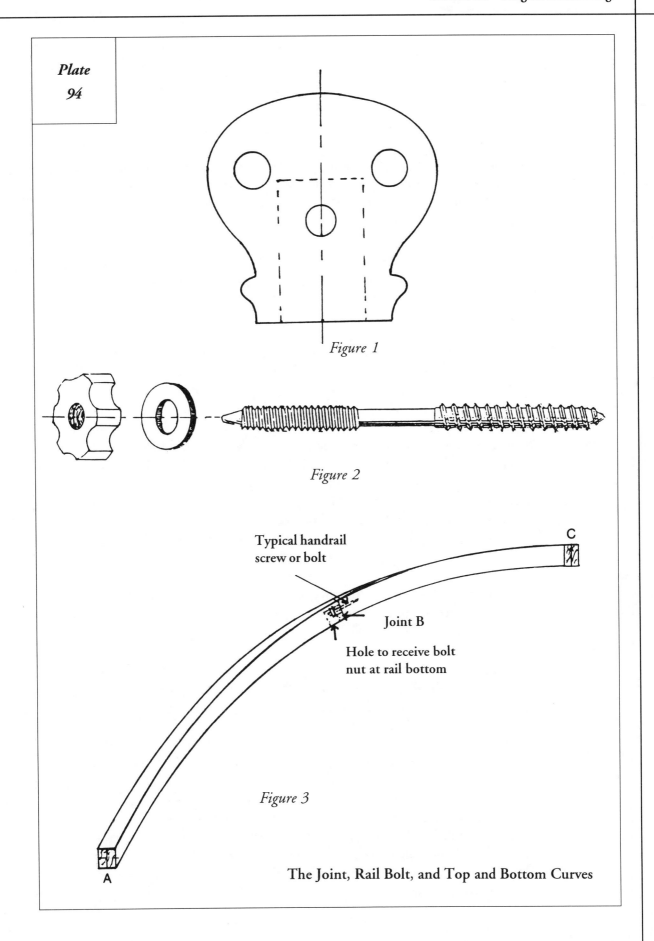

Plate 94

Figure 1

Figure 2

Typical handrail
screw or bolt

C

Joint B

Hole to receive bolt
nut at rail bottom

Figure 3

A

The Joint, Rail Bolt, and Top and Bottom Curves

Plate 95—Curvature of the Rail Sides through the Joint

Whenever two incline-turn handrail sections are to be joined together at the spring line, the width of the rail at the joint should always be left slightly wider than the normal rail width. By doing so, the projection of the top half of the upper section and the bottom half of the lower section into each other's curve can be worked to a smooth transition.

This projection of one section into another at the joint becomes considerable whenever the plan radius is short, the pitch is steep, and the rail is deeper than it is wide. Whenever this occurs, the sides of the rail at the butt joint will show a decided curve.

Figure 1—Let us assume the radius of this half-circle plan is 9" to the rail center. The handrail size is to be 2" wide and 7" deep. Two rail sections will be required to make the half-circle climbing-turn from A to E. The plan tangents are AB, BC, CD, and DE.

Figure 2—The tangents of both sections are unfolded and shown equally pitched from F to H. The total height of both sections is E'H. The joint of the two sections is at G. Since both sections are equal, one layout to find the face mold will serve for both. CG and JH are equal heights. To lay out the upper section only, GJ will be the horizontal base line. Draw half of the rail thickness at each side of the pitched tangents at joint G. Draw the joint TT' through G.

The shaded areas show the upper half of the upper section and the lower half of the lower section projecting into the curve of the adjoining section. This is also shown in **Figure 1** as UU'.

Figure 3—The face mold is found in the same manner as in **Plate 73** by either using ordinate parallels drawn at the plan being transferred to the oblique plane or by using the trammel method. Use of the trammel in such a short radius and compact area would be the better choice because fewer layout lines are required. The spring lines at the face mold are at K and H. Since straight rail is to be connected to the top joint, the face mold includes a straight section beyond H as HM. M is located so that the top half of the rail (shown dotted) will extend past the vertical extension of the spring line. Although the joint between the two curved sections is at K, the face mold curves at this joint should also include the portion of rail projecting into the curve of the adjoining curved section. By doing so, the surfaces of the rail block can be properly marked to indicate the excess stock to be removed to form the sides, as shown in **Figure 4**. Block thickness will determine the

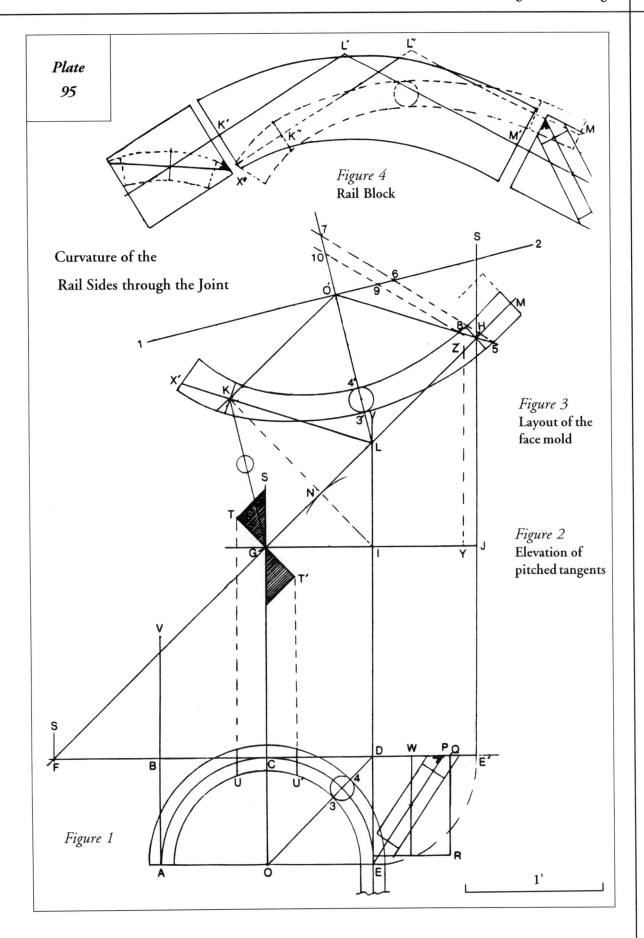

Plate
95

Figure 4
Rail Block

Curvature of the
Rail Sides through the Joint

Figure 3
Layout of the
face mold

Figure 2
Elevation of
pitched tangents

Figure 1

1'

distance the face mold must slide sideways along the block surface in order to meet the new tangent lines indicated by the bevels. An additional amount of face mold curve must then be added to K in order to reach the butt joint. Generally, 4" to 6" beyond K is more than ample to cover the steepest pitch, sharpest curve, and thickest handrail likely to be encountered.

Therefore, in **Figure 2,** using the trammel method to find the face mold, JY equals half of the plan rail width and H-5 and H-8 equal HZ. L-4'-3'-O' in **Figure 3** equals D-4-3-O in **Figure 1.**

Perpendicular to the minor axis LO' in **Figure 3,** draw the major axis at O' as 1-2. For the inside curve trammel-rod lengths, let 8-9 equal 3'-0", and extend to 10 for rod lengths 8-9-10. For the outside curve trammel-rod lengths, let 5-6 equal 4'-0" , and extend to 7 for rod lengths 5-6-7.

Figure 4—With the bevel drawn in **Figure 1,** showing DF to be equal to arc IN in **Figure 2,** DO and QR are shown to be the block width and thickness. This is again shown in the butt joint view in **Figure 4.** Although the face mold curves are correct, the block size is not the same, but larger. This is determined by the bevel and the depth of the handrail. This condition is shown in **Plates 95** and **96.** Sliding the face mold in **Figure 4** shows that the extra curve of the face mold from K" to X" is necessary in order to show the exact stock to be removed to properly dress the sides. Once the sides have been dressed to the face mold curves, the curvature of the rail will be shown to occur across the face of the butt joint.

Plate 96—The Rail Block and Face Mold Joint Widths and Block Thickness

Figures 1 through **5** show butt joints of handrail sections inclining from left to right. D represents half the face mold width as determined by the bevel required for a joint. A represents the block thickness in all figures.

Figure 1 shows the butt joints of both lower joint B and upper joint C of an arbitrary rail section requiring the same bevel. The rectangular size of the rail at the joints shows that the block width should be slightly wider than the face mold in order to allow more surface on which to mark the face mold curves.

Figures 2 and **3** represent arbitrary sections with a greater bevel at one joint than the other. At joint B, the bevel is such that the block width required at this joint is less than the width of the face mold joint. Whereas at joint C, the block width must be greater than the width of the face mold joint.

Plate 96

The Rail Block, Face Mold Widths, and Block Thickness

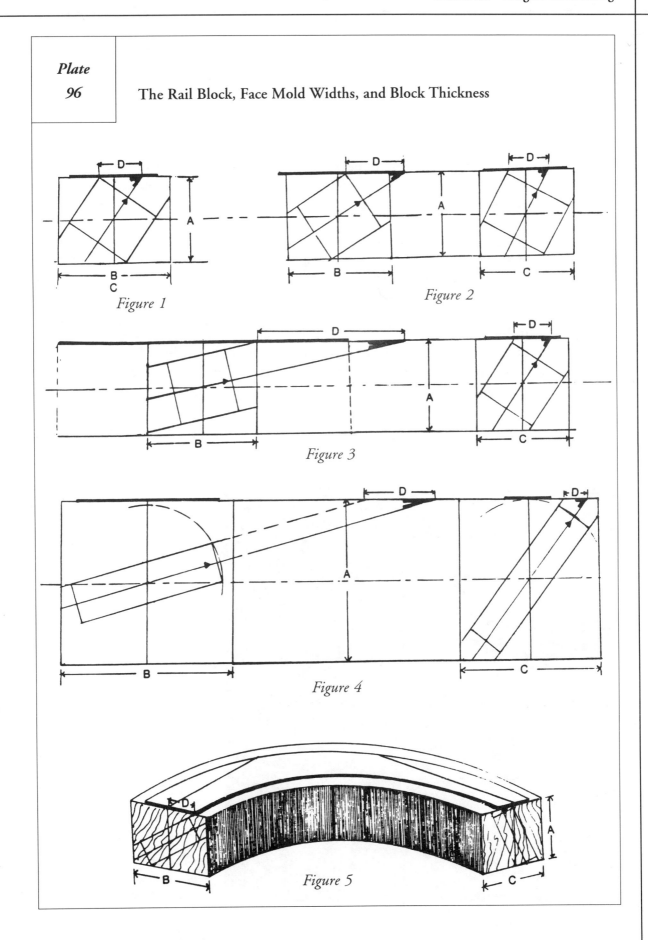

Figure 1

Figure 2

Figure 3

Figure 4

Figure 5

Figure 4—The bevels are similar to those in **Figure 3**, but the rail size is considerably different. The narrow rail width and great depth shows that the block width at joint B must be slightly wider than the face mold joint width. At joint C, as at B, the block width must be wider than the width of the face mold joint.

Figure 5 is a perspective view of a rail block for a rail similar to **Figure 4**. The exception to the block thickness in **Figure 4** is that this block can be thinner because the bevels are applied from the same side, which does not require the rail to twist from one side to the other.

It should be pointed out at this time that the likelihood of such a great difference in bevels occurring in **Figure 3** and 4 is improbable. However, the bevels are purposely shown in this manner to illustrate that the width of the face mold joints are not necessarily that of the rail block joints.

This plate shows the importance of first drawing the rectangular size of the rail at the bevels prior to getting out the rail block thickness and width.

Glued stock from the same plank for making solid, twisted handrail sections.

Plate

97

Plate 97—Optional Method of Finding the Face Mold Whenever One Tangent Is Level

This is a combination of using both methods "A" and "B" principles to find the shape of the face mold. AB and BC are the plan tangents. AB is the level tangent of an obtuse turn, and B'D is the pitched tangent to gain the height C'D.

The layout to find the angle of the elevated tangents is done by using method "A", the same as in **Plate 81**. The bevels are also found in **Plate 81**.

To find the shape of the face mold, parallels to ordinate AB, intersecting plan tangent BC can be transferred to the pitched tangent B'D from where they can are drawn parallel to EB' the same length as they are from BC to the plan curves.

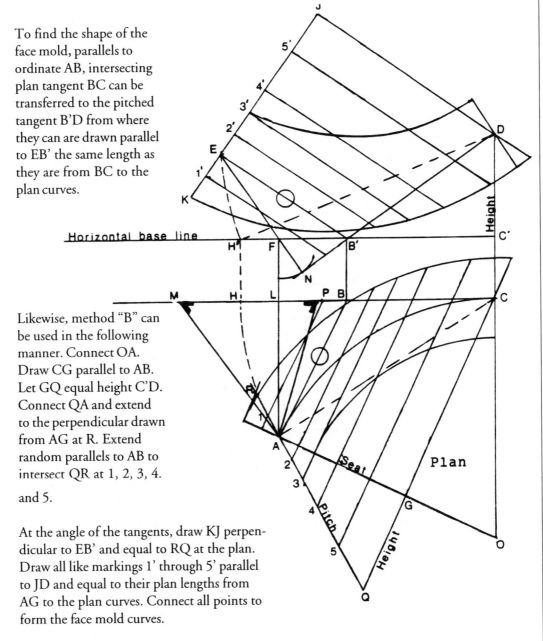

Likewise, method "B" can be used in the following manner. Connect OA. Draw CG parallel to AB. Let GQ equal height C'D. Connect QA and extend to the perpendicular drawn from AG at R. Extend random parallels to AB to intersect QR at 1, 2, 3, 4. and 5.

At the angle of the tangents, draw KJ perpendicular to EB' and equal to RQ at the plan. Draw all like markings 1' through 5' parallel to JD and equal to their plan lengths from AG to the plan curves. Connect all points to form the face mold curves.

Plate 98—Increasing Tangent Lengths Whenever the Radius Is Small

Prior layouts have shown the plan tangent lengths to be from the spring lines to the vertex. As already stated, this is not a fixed rule. Plan tangent lengths may extend beyond the spring lines if it would be better to make a more sizable or workable layout. **Plate 98** is such a plan.

Whenever the distance from the vertex to the spring lines is extremely short, as in this obtuse plan, making the layout within the cramped area of the short tangents does not make much sense when the same layout can be made large enough to be more easily worked. Therefore, in this particular plan of a turnout where there is a level and rake tangent, the tangent lengths can arbitrarily be AB and BC.

This layout is conveniently drawn at the plan by the method "B" principle shown in **Plate 81** rather than by method "A". CK is the ordinate direction line, parallel to the level tangent AB.

Seat KA is drawn perpendicular to KC through A and extended to M, MK being the parallel distance between the width of the turnout and CK. KL is equal to height C'D. Connect LA for the pitch. Draw NM parallel to LA. Random parallels drawn at the plan turnout are extended to the pitch NM, then perpendicular to NM to find the angle of the tangents and the face mold.

The bevels are found in the two manners already shown in methods "A" and "B". In method "A", the bevel for the pitched tangent is AHF and is found by making FH equal to arc F'G. The bevel for the level tangent is found by making FI equal to height C'D for the bevel AIF.

These same bevels are also found in the method "B" manner as shown. Along either the pitch line LA, or its parallel NM, draw a perpendicular line to intersect the height line, say from S to K. Then, from N draw parallel lines to the tangents RQ and RP. NP is already parallel to RQ, but NU must be drawn parallel to PR. Now, from S as the radius point, strike an arc touching NU to intersect at T. Connect TK for the bevel KTM for the pitched tangent joint P. An arc from point S touching parallel PN at N already shows N connected to K, therefore KNM is the bevel required for the level tangent joint Q. If the bevels were to be found from pitch line LA, with S' the radius point, they would still be the same as those already found.

Certainly, the bevels found in the method "B" manner are accurate, but it is obviously much simpler to find them along the line of the stretched out tangents as in **Plate 81**.

Plate
98

Increased Tangent Lengths whenever Radius Is Small

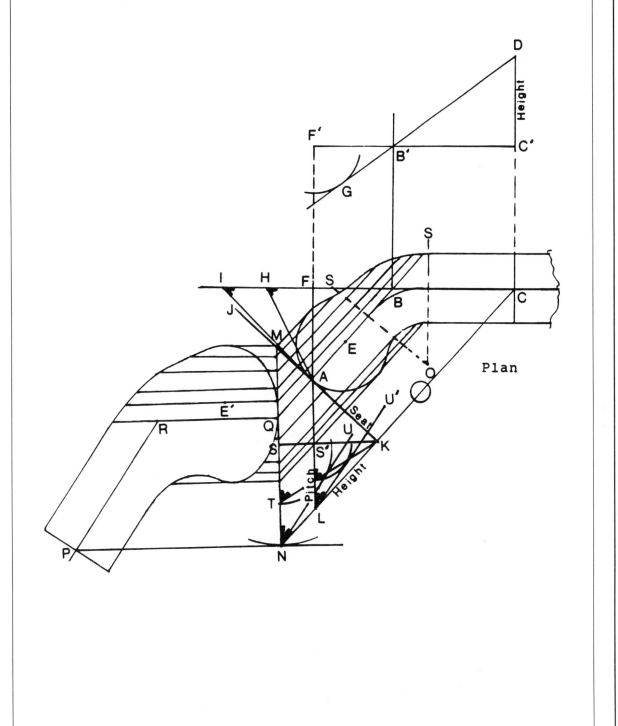

Plan

Plate 99   Obtuse Plan with Gooseneck

The tangent pitches in this plate are similar to those in **Plate 79** except that in this plate the pitch of the upper tangent strikes exactly at H, the same as in **Figure 6** of **Plate 67**. Consequently, no bevel is required for the top joint. The surface of the top of the rail block must extend beyond the spring line D to point L so that the bottom curve of the gooseneck will be within the block's top surface. The block width at this upper tangent will be level across so that once the sides have been dressed, additional material can be glued to the surface at K to allow for the gooseneck. The bevel shown at D or E is the vertical line of the gooseneck and is applied from the bottom surface of the rail block.

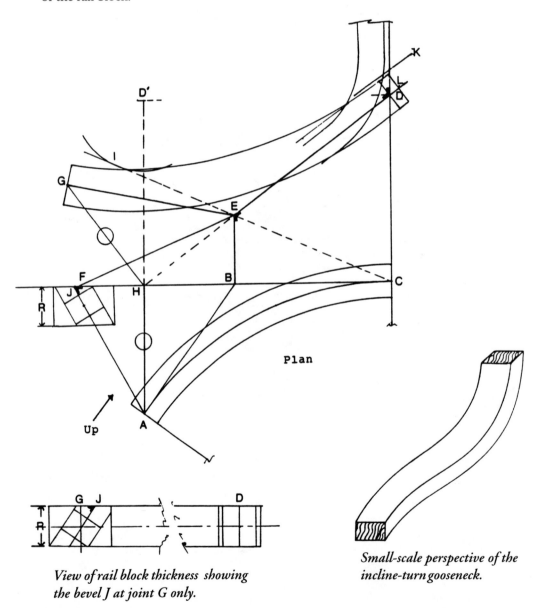

Plan

View of rail block thickness showing the bevel J at joint G only.

Small-scale perspective of the incline-turn gooseneck.

Plate 100

Obtuse Plan with Lower Overeasement

The tangent pitches of this plate are similar to those in **Plate 80** except the lower tangent is pitched so that when C is drawn through E it will strike the height HD' exactly at D', as shown in **Figure 5** of **Plate 67**. Consequently, no bevel will be required at the lower joint G. As in **Plate 99**, the lower joint should extend past spring line G so that the top curve of the overeasement will be within the butt joint of the block surface M at L. The block width along the line of the lower tangent will be level across. Once the sides of the rail have been dressed, additional material can be glued to the bottom surface to form the desired overeasement. N is the plumb bevel as in **Plate 99**.

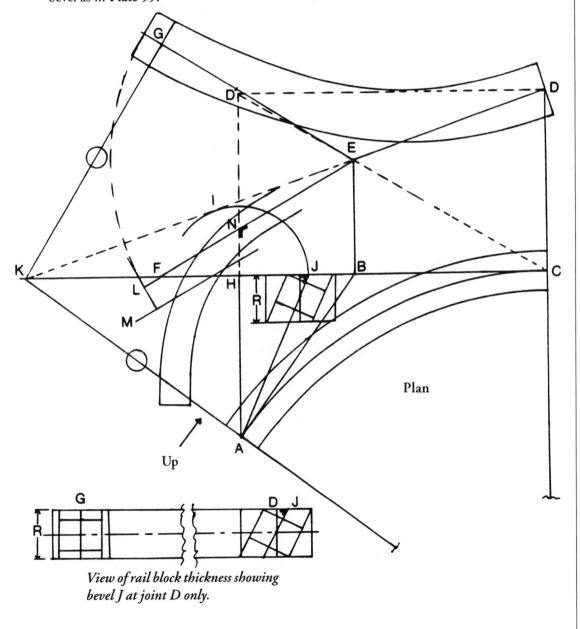

Plan

Up

View of rail block thickness showing bevel J at joint D only.

Plate 101

A Turn Section at Platform to Include Two Small-Radii

Included in this plan of a quarter-turn handrail section around a platform corner are two small-radius turns. Rather than make two separate sections with a joint at X, it is much more practical to make one section to include both small turns with some straight rail added beyond spring lines Y'. The joints can be at any point, say at A and C. The plan tangents for the layout are then AB and BC. The face mold layout is the same as in **Plate 73**.

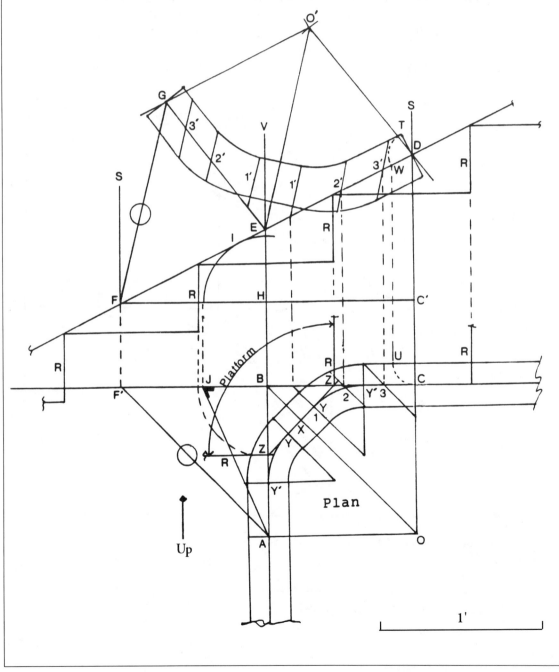

Plan

Up

1'

View through balcony balusters of a continuous handrail around a platform turn.

Plate 102—Forced Ramp Within the Rail Section

Figure 1 is a quarter-circle plan of an incline-turn handrail section inclining from left to right. Assume there are square treads below the riser indicated as 1 and above the riser indicated as 4. The total dimension between risers 1 and 4, along the straight handrail centerline and the plan tangents, is greater than three dimensions between risers of a square tread. Consequently, the normal pitch over both lower and upper square treads will not be in the same line. This alignment problem can be overcome by making the adjustment within the turn section itself rather than by making a separate easement beyond a spring line (see **Plate 133**).

Let the centerline pitch over the lower square treads and both pitched tangents be in the same line. The difference between the parallel pitches over both lower and upper square treads is shown to be at K between risers 4 and 5. The layout to make the face mold for this turn can be accomplished with equally pitched tangents either by making the adjustment at the upper tangent joint or by making equal adjustments at both joints. In this layout I have chosen to show the adjustment made at the upper joint only.

The turn section is to consist of equally pitched tangents FE and ED. The horizontal base line is CF. Straight rail is extended beyond both spring line points of the face mold, as at R and G.

While joint R will remain in the same position along the centerline at the lower joint, G must be raised plumb along the applied bevel on the joint so as to meet the normal straight rail centerline at L. For the bevel, let BI equal arc BH. I, connected to A, shows the bevel as AIB. With the bevel marked at the base line FC, draw half the width of the rail at both sides, and show the rectangular size of the rail with its center L'. Mark the distance from L' to G' the same as K or LG shown at the upper spring line elevation. Again, with G' the rail center, draw the rectangular size of the rail. From G', draw QM perpendicular to FC. The block thickness is shown to be PU and the block width UX. Let QM represent the perpendicular line drawn from the block surface at both joints of the face mold tangents prior to sliding the face mold. Although not shown in this plate, the face mold is found as in **Plate 73**. Half the width of the face mold joints is shown as IQ at the bevel. As in **Plate 95**, the block width is shown to be much greater than the face mold width.

Figure 2 is a side perspective of the rail block showing the joints in elevation. The block width at both joints is U'X', the same as UX in **Figure 1**. The block thickness is Q'M, and the block length is the same as the face mold.

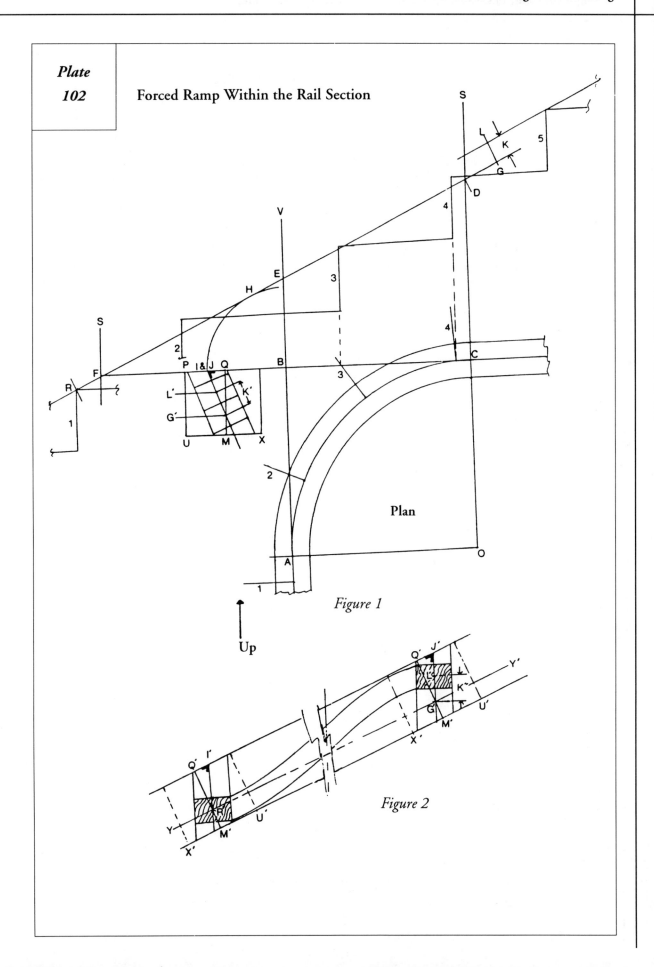

Plate 102   Forced Ramp Within the Rail Section

Figure 1

Up

Figure 2

Square the joints to the block surface and to the tangents at the joint. Mark the tangents on the top of the block, then square across the surface of the butt joints as Q'M' and mark the tangents on the bottom surface at M.

Mark the distance UG' of **Figure 1** as dotted margin YY" in **Figure 2**, intersecting Q'M' at R' and G'. Apply the bevel I of **Figure 1** from the top of the block through centers R' and G'. Draw half the width of the rail parallel to the bevel lines. R' is the normal centerline of the rail at the lower joint. Whereas, the upper normal centerline at G' must be raised the distance K, or G'L' of **Figure 1**, as K" or G"L" in **Figure 2** in order to establish the correct centerline of the upper joint as L".

With the face mold shifted at the new tangent positions drawn on the block surfaces through the bevels, the stock to be removed from the rail sides is marked with the face mold, and then dressed accordingly. Once the sides have been dressed, at the center of the block length, mark the center of the block thickness. Connect the section to the adjoining sections, if any. Draw the transition curve of either the top of bottom of the rail from the adjoining sections through either the top or bottom location at the center of the block length. Disassemble and dress the curved ramp marked on the section. Scribe to thickness, and work the opposite side.

Handrail craftsman "squaring" a right angle twist of a tangent handrail section, before molding.

Plate 103—Face Mold Cannot Be Used in Some Rake-to-Level Obtuse Plans

Normally, the face mold is used to indicate the stock to be removed in order to form the sides of an incline-turn handrail section. In rake-to-level obtuse sections, the bevels will be applied to the joints from the same side or edge of the block. (See **Figures 2, Plates 67** and **69**). Because the bevels are usually very acute and are applied from the same side of the block, the rail will not have a great twist through the turn. Consequently, the rail block need only be minimal width. Therefore, the face mold does not have to be found since it cannot be applied to the block surface. **Figure 1** is such a plan. AB and BC are the plan tangents. ABCD is the plan parallelogram. AOC is the centerline curve. **Figure 2** shows the horizontal base line EF extended. EI is the height. FI is the pitched tangent. From **Figure 1**, extend AG to J in **Figure 2**. JK is drawn perpendicular to tangent FI. FN equals level tangent AB in **Figure 1**. Connect NF for the angle of the elevated tangents NFI.

For the bevels in **Figure 1**, draw perpendiculars A'H and A"H" at any convenient location equal to AG. For the level joint bevel, let HM equal height EI in **Figure 2** and connect to A'. For the rake joint bevel, let H"L equal arc JK in **Figure 2** and connect to A". Draw the rectangular size of the rail at both bevels.

Because the bevels are applied from the same side of the block, the block width 2 in **Figure 1** will be 2' in **Figure 2**. Although the face mold is not required and does not have to be found, I have shown it to be UVWX in **Figure 2**. (see **Plate 81**).

To find the rail centerline at T in **Figure 2**, first draw BOD in **Figure 1**. From O draw a parallel to AB, the distance between being P. Transfer P as the parallel width Q along the bevel line MA', intersecting MH at R. Transfer MR to **Figure 2** as NS (which coincidentally happens to be half of 2'). Draw ST parallel to NF and equal in length to the parallel at O in **Figure 1**. With position T located in **Figure 2**, the centerline curve can be freely drawn through N, T, and I. Half of the dimension 2 is then marked at each side of the tangent at the joint, and the inside and outside curves drawn reasonably parallel to the centerline curve for the width of the block. The block thickness is shown as dimension 3 in **Figure 1**.

The rail block is then made to the minimum width of the pattern shown in **Figure 2**. Joints N and I are dressed square to both the tangents and block surface. The tangents are drawn perpendicular from the block surface through the center of the block at the butt joints. The bevels are applied

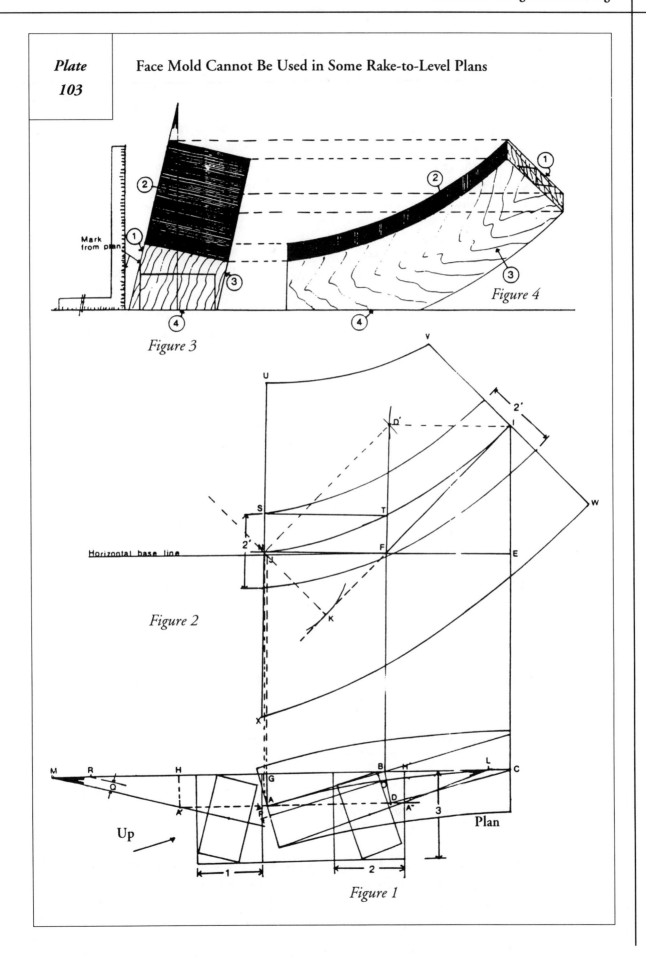

Plate 103

Face Mold Cannot Be Used in Some Rake-to-Level Plans

Figure 4

Figure 3

Mark from plan

Horizontal base line

Figure 2

Up

Plan

Figure 1

through the block center, and the rectangular size of the rail is drawn, as shown in **Figure 1.** In **Figure 3,** the bottom of the rail at the level joint is dressed square to the joint face so as to create a seat for the block so that it can be placed over the plan drawing with the bevel lines falling directly over the plan-drawn tangents.

With the block in position over the plan, the outside (convex side) curve can be transferred from the plan to the block with a carpenter's framing square as shown. With the squared seat resting on the bandsaw table, and the height of the block within the dimensions of the bandsaw, the curve thus marked can be cut plumb. If the height of the bandsaw opening does not permit the block to be cut plumb, then the side can be roughly cut at the saw and then dressed plumb using either hand tools or abrasive cutting. The circled numbers in **Figure 3** and in the side view in **Figure 4** represent the same portion of the rail block.

Application of the face mold UVWX is not practical because the surface of the block will not show the shifted, plumbline tangents determined by the bevels, which are necessary in order to slide the face mold. Therefore, using the framing square at the plan is the practical method of making the outside curve.

Plate 104—One Rail Section for Two Stairways Meeting at the Top Landing

Figure 1 shows the two top risers of straight runs of identical stairways (one right-hand and one left-hand) meeting at the top of both stairs. The rake-to-level handrail section shown will meet the straight railing of both stairs. Assume the curve of the rake-to-level rail section is as shown. The heights are to be 2'-8" over the risers, and 3'-0" at the floor level.

The top risers of both stairs must be located at the plan so that the rake and level tangents of the turn section meet the vertex at the proper height. First; with spring lines A and E established at the plan, divide the curve into four equal length tangents, or two equal rail sections, ABC and CDE. The normal joint of the two sections at C is to be eliminated, making one section encompass the entire turn.

Figure 2—Stretch out the plan tangents in **Figure 1** and extend the spring lines and vertices as vertical lines. At any convenient height, draw a horizon-

Plate 104

One Rail Section for Adjoining Stairs

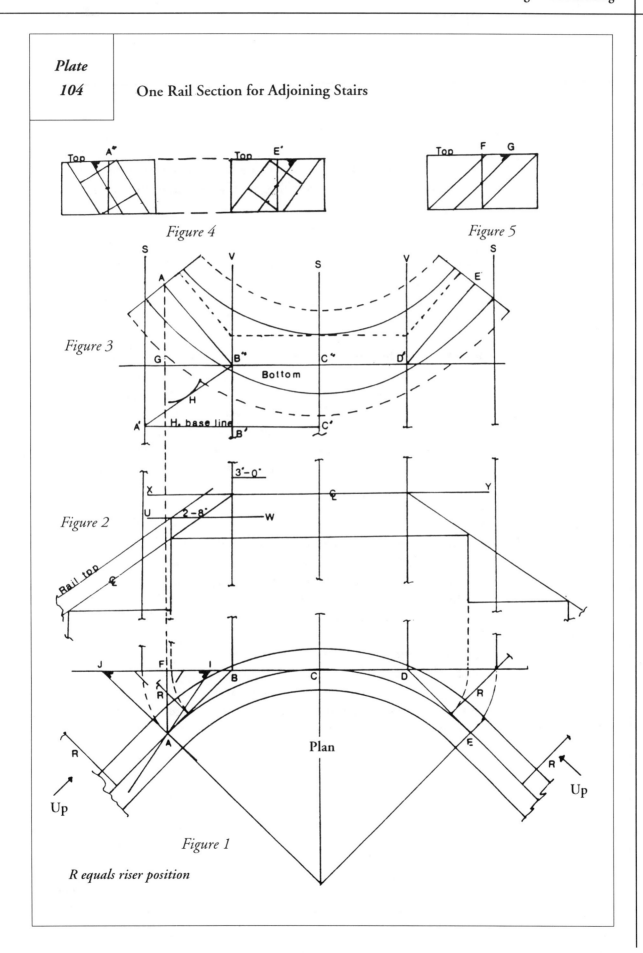

Figure 4

Figure 5

Top A"

Top E'

Top F G

S V S V S

Figure 3

A

G B" C" D' E'

Bottom

H

A' H, base line B' C'

Figure 2

X 3'-0" Ȼ Y

U 2-8' W

Rail top Ȼ

J F I B C D E

R R

A Plan E

R R

Up Up

Figure 1

R equals riser position

tal line through the vertical spring lines and vertex line, such as XY to represent the level rail centerline. Mark half the rail thickness above the centerline as the top of the level rail 3'-0" above vertex line B. At the intersection of the level tangent at vertex B, draw the normal pitch of a tread and riser. This will be the rake tangent pitch.

Assuming the rake rail height is 2'-8' over the risers, mark down 4" from the 3'-0" height at the vertex. This will establish the top of the rake rail over the top riser as UW. With the position of the top riser now established at the left-hand stair, transfer it to the right-hand stair and then to the plan in **Figure 1** as R on both sides.

Figure 3—The layout for the rake-to-level obtuse plan has already been shown in **Plate 82** and is repeated here. Once the face mold is completed for the left-hand stair, simply turn it over at joint C for the face mold for the right-hand stair. Together, this will be the face mold section for both stairs. The bevels in **Figure 1** are obtained by making FI equal to arc GH in **Figure 3** for bevel FIA for the pitched tangent joints at A" and E'. The bevel at C" obviously cannot be applied, however, it must be found. Make FJ in **Figure 1** equal the height C'C" in **Figure 3** for bevel AJF.

Figures 4 and **5**—The bevel FIA in **Figure 1** is shown applied in the proper manner for joint A" in **Figure 4**, and reversed for the joint at E'. In **Figure 3**, since bevel AJF of **Figure 1** cannot be applied at C", tangents B"C" and C"D' are simply shifted on the rail block as shown at **Figure 5**. **Figure 5** shows the shifting of the face mold the distance FG when the new tangent lines, indicated by the simulated application of the bevels, are drawn.

The dotted tangent lines in **Figure 3** show the shifting of the tangents on the surface of the block. The shifting of FG in **Figure 5** will be toward the convex side of the block at the top surface, and toward the concave side at the bottom surface. Block width, indicated by the bevels at **Figure 4**, is shown by the dotted curves at **Figure 3**, allowing ample surface to mark the face mold.

Plate
105

Increased Block Thickness When Tangents Are Unequally Pitched

This is an acute plan of a handrail section showing unequally pitched tangents in elevation. A stretch out ramp of the rail thickness at the concave side of the rail is shown from G to H. The stretch out of the concave side at the plan is 1-2 and is marked at the base line as CG. AB and BC are the plan tangents. I and H are the joints. GI' equals FI. The rail thickness is marked at each side of I'J and EH. The ramp easing, in this instance, is freely drawn to suit. The bevel at M shows the block size to be N. With a line drawn from the bottom of the upper joint to the rail bottom at I', the width at P is marked. The block thickness should be a minimum thickness of N or P, whichever is the greatest.

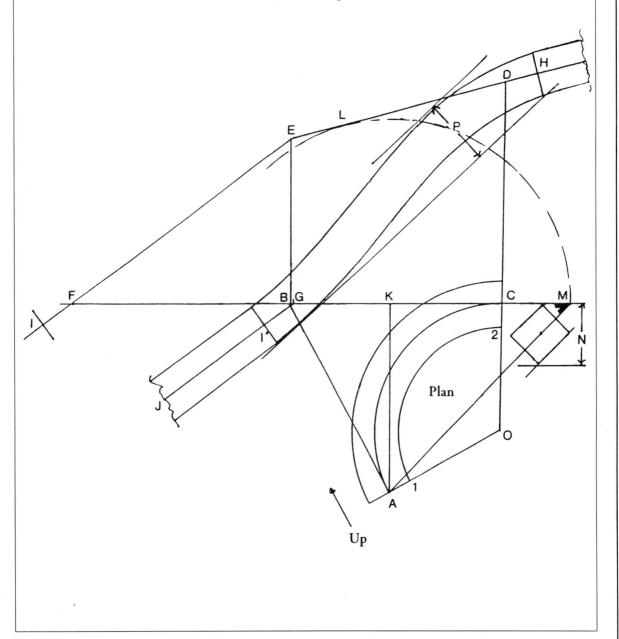

Plan

Up

Plate 106—Quarter-Circle Plan Showing How to Find
the Baluster Lengths from the Rail Center to the Horizontal Base Line
or to the Tops of Treads

Figure 1—Assume that the baluster centers are at <u>a</u>, <u>b</u>, and <u>c</u> at the plan. From the balusters' centers draw parallel lines to the ordinate BO to intersect the plan tangents at <u>a</u>, <u>b</u>, and <u>c</u> . Transfer <u>a</u>, <u>b</u>, and <u>c</u> to the horizontal base line and then to the pitched tangents as <u>a'</u>, <u>b'</u>, and <u>c'</u>. The heights from <u>a'</u>, <u>b'</u>, and <u>c'</u> to the base line along A'C are also the same heights from <u>a</u>, <u>b</u>, and <u>c</u> at the plan curve to the centerline curve along the oblique plane.

Figure 2 is a prismatic view of the plan and pitched tangents related to the centerline curve. Since <u>a'a"</u>, <u>b'b"</u>, and <u>c'c"</u> are parallel to <u>aa,</u> <u>bb,</u> and <u>cc</u> at the plan, it proves that the baluster heights found at the pitched tangents are also the same heights to the centerline curve along the oblique plane.

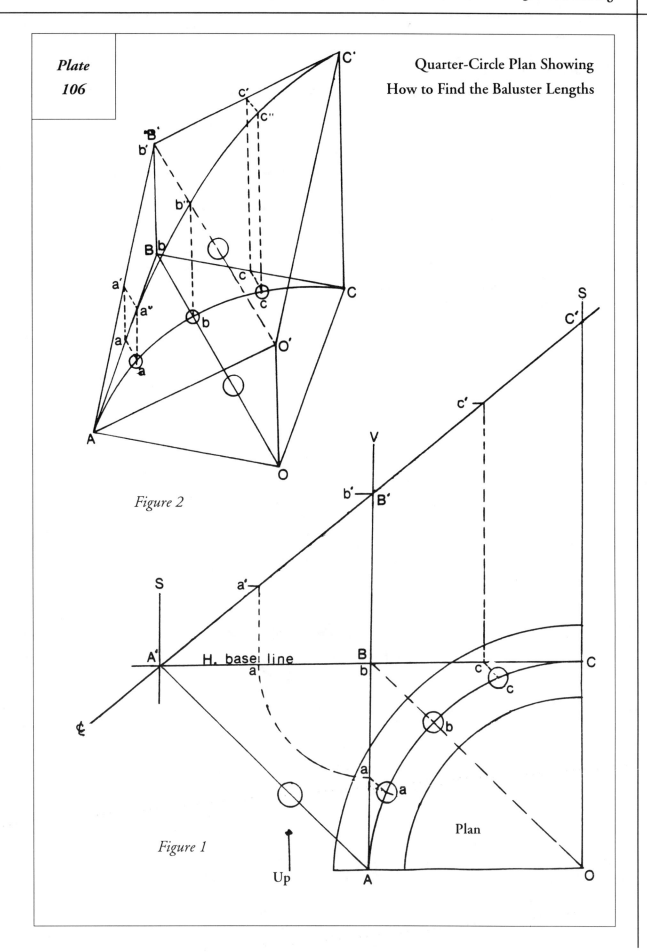

Plate
106

Quarter-Circle Plan Showing
How to Find the Baluster Lengths

Figure 2

Figure 1

Plate 107	Top and Bottom of Squared Rail Are Level to Radius

Since the sides of the squared rail will fall plumb over the sides of the rail at the plan, it then follows that the top and bottom surfaces of the rail, at any point throughout the turn, must also be level to the radius swung from the imaginary vertical extension of the radius point from the plan.

This plate shows an isometric view of a quarter-turn squared handrail section inclining from A to C. OO' is the imaginary vertical extension of the radius center point O. XY is an arbitrary level-to-radius location along the top surface as swung from OO'.

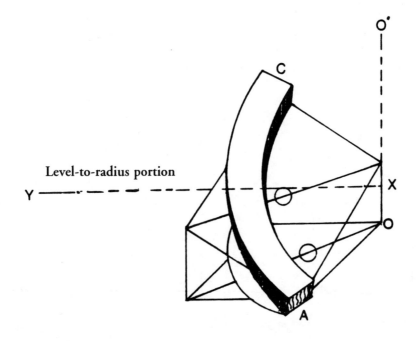

Level-to-radius portion

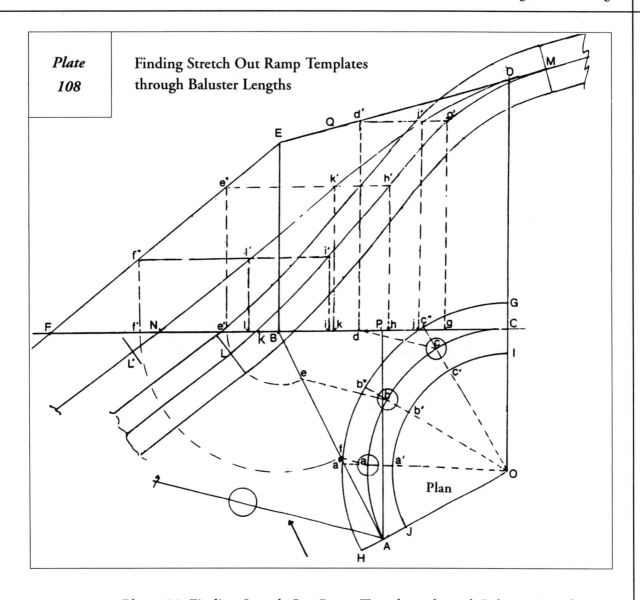

| Plate 108 | Finding Stretch Out Ramp Templates through Baluster Lengths |

Plate 108–Finding Stretch Out Ramp Templates through Baluster Lengths

Figure 1 It has been established in **Plate 106** that in order to find baluster heights at the rail centerline the balusters must first be located at the pitched tangents. It is also found, in **Plate 107**, that the heights at the center of the rail width are also the same heights at the centerline of the rail thickness at each side of the rail width along the line of the imaginary level radius. Therefore, baluster heights at both sides of the rail width are equal to those heights found at the pitched tangents. For either side ramp, baluster centers $\underline{a}$, $\underline{b}$, and $\underline{c}$ at the plan are drawn parallel to the ordinate to intersect the plan tangents, from where they are transferred to the pitched tangents as $\underline{d}'$, $\underline{e}''$, and $\underline{f}''$. The centerline stretch out of each ramp is then drawn through the baluster heights that have been established along the stretch outs of the plan curves.

For the inside ramp: CK equals the stretch out of the curve IJ. $\underline{g}$, $\underline{h}$, and $\underline{i}$, along CK, equal the stretch out positions of $\underline{c}'$, $\underline{b}'$, and $\underline{a}'$ at the plan. From

K, extend several inches of straight rail as KL the same pitch as EF. At D, let length DM equal KL. Set up heights g-g' as d-d', h-h' as e'-e", and i-i' as f-f". Connect points L, K, i', h', g', and M for the inside ramp centerline. Set off half of the rail thickness at each side of joints L and M and draw the ramp. For the outside-radius side ramp: Let CN equal GH at the plan. j, k, and l equal the stretch out positions of c", b", and a". Let NL' equal KL, j-j' equal d-d', k-k' equal e'-e", and l-l' equal f'-f". Draw the centerline ramp curve L', l', k' j', and M.

Figure 2—Although the stretch out curves of the plan rail width are shown as CK and CN, along line FC in **Figure 1**, they can also be drawn separately at any convenient area; the total height, and heights at the tangents being the same as in **Figure 1**. **Figure 2** shows the stretch out ramp of the convex side of the rail. All lettering equals that of **Figure 1**.

The joint-to-joint transition curves of any incline-turn handrail section must be as pleasing to the eye as possible. In a short-radius incline-turn section where there are no adjoining turn sections, the joints may he extended well beyond the spring lines in order to allow for a more graceful transition ramp to be drawn between the joints along the concave side of the rail rather than if the ramp was confined between spring lines. These more pleasing lines may cause the rail to be slightly canted in width throughout the turn instead of level to width as shown in **Plate 107**. However, slightly canting the width of the rail will not adversely affect the appearance of the rail nor the tooling of the rail profile.

Plate 109–Finding the Steep Pitch of Balusters at the Rail Bottom.

In **Plate 108** the centerline pitches of both side templates are obviously different. Whenever square-top balusters are used, the pitch of the cut at both sides is also different. It is the steeper pitched side that must be found if the squared-top lengths are to be uniform. In the quarter-circle plan of equally pitched tangents in **Plate 109**, the centerline ramp of the rail thickness is at the concave side of the baluster width, as at DE in the plan. CF equals stretch out DE. With baluster center heights found as in **Plate 108**, the transition curve is from G to H. Half of the rail thickness is then drawn below for the true stretch out pitched cut of the balusters. The top square portion of turned balusters can then be made uniform, or as desired.

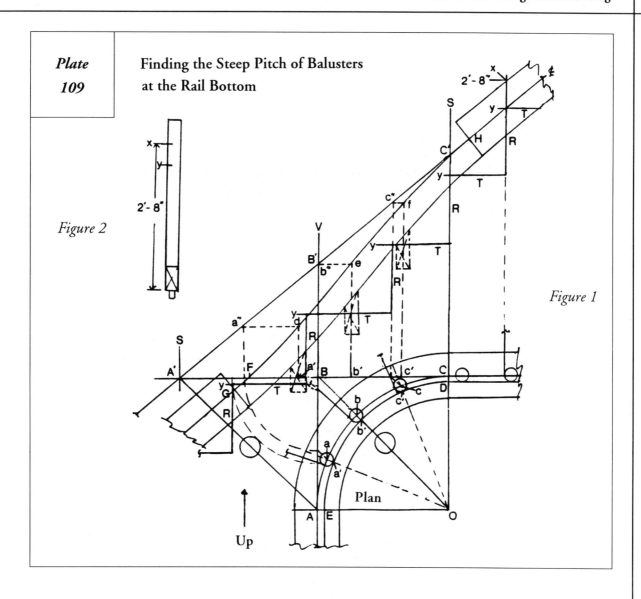

Plate 109

Finding the Steep Pitch of Balusters
at the Rail Bottom

Figure 2

Figure 1

Plan

Up

To find the baluster lengths at each tread, with R and T identifying riser and tread lines, let 2'-8" be the desired height over a normal square tread, as at x in **Figure 1** elevation. Mark y as the tread line. In **Figure 2**, take a thin flat rod the same width as a baluster, approximately 3'-6" in length, and mark the 2'-8" height from the baluster bottom as x. Take the rod and place the x mark over the x at the 2'-8" height in **Figure 1**, and mark position y on the rod. Now, with y on the rod placed on each tread at the baluster location, the exact cut of balusters can be marked.

Baluster lengths and pitches are based upon using the side ramps found as in **Plate 108**. Baluster cuts are then made at the installation of the handrail.

Plate 110—A Forced Ramp to Meet Different Outside Pitches

It has been shown how an easement is made within the turn section whenever two equally pitched centerline pitches beyond the spring lines do not meet at the vertex. In this quarter-turn plan the rail pitch beyond both spring lines is different. This layout will show how a forced easement can be made within the turn section in order to bring two differently pitched centerlines into alignment. The centerline pitches beyond the spring lines in this plan are GJ and HI.

Rather than having the plan tangents terminate at the spring lines, let them extend beyond the spring lines as BA and BC to allow more freedom to make the transition ramp at the concave side of the rail section. Tangent BA is stretched out as BF. The tangents are shown to be equally pitched along the line of JG to intersect the upper straight rail pitch at H. The joints will be at G and H. While the straight rail JG will have a square cut at G to butt into the block joint, the straight rail at H will have to be cut on a bevel, 1-2, to meet the butt joint of the rail block. This will be much more convenient than cutting the rail block to meet the square joint of the straight rail.

The ramp for the concave side is found by stretching out KL as CM. Pitch MP equals GJ. With the joints at N and H, draw the ramp to suit. If there are baluster lengths and pitches that must be found, then they are found, as in **Plate 109,** with height adjustments made to suit the freely drawn ramp. Block thickness is increased to a minimum of that shown at Q because of the abrupt overeasement at the top of the rail.

Plate 110   A Forced Ramp to Meet Different Outside Pitches

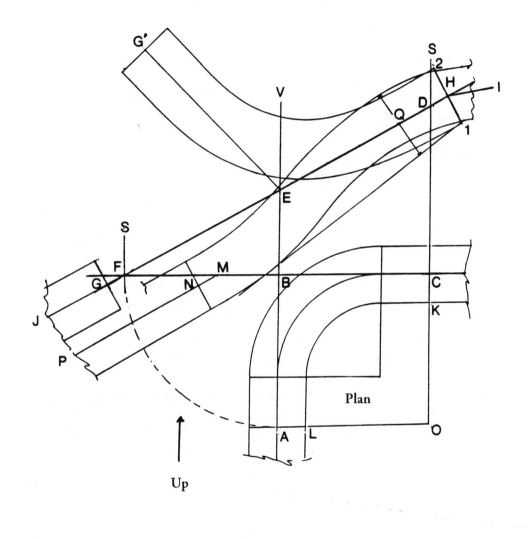

Plan

Up

Plate 111 —Stretch Out of Treads and Risers at Rail Centerline Drawn from a Straight Edged Board or Straight Line

It is not always possible or practical to stretch out the treads and risers at the plan, but it could be done from a straight edged board, or from any straight line drawn at a convenient area. **Plate 111** shows how this is accomplished.

Figure 1 is a plan of a three-winder stairway with a wall-type handrail at both inside and outside corners. The riser height is 7-½". Mark down the dimension between square-tread risers 9 through 12 along the centerline of the handrail, including the tangents, plus the 10" width of one square tread. Divide the total dimension by 4 (number of treads in the dimension). The resulting figure will be the average tread width to be used in the stretch out, the riser remaining a constant 7-½". The average tread width for the stretch out tread for the inside rail centerline is found to be 6-¼"; for the outside rail centerline, 25-¾".

Figure 2 shows two drawings of a framing square inserted in the slots of a straight piece of wood stock, say 1-½" thick, 2" wide, and 3' in length. The stock is center-slotted in thickness approximately 14" at each end to receive the framing square. The square is set to both inside and outside rail pitches and secured in place with carriage bolts and wing nuts as shown.

Figures 3 and 4—With straight lines drawn to represent the edges of straight-edge boards, let bevels C in **Figures 3** and 4 represent the average pitches at riser 13 in both stretch outs. By sliding the pitched framing square along the edge of the board, risers 12 through 8 are easily drawn according to their dimensions at the plan. The plumblines of the vertex and spring lines are also drawn. Start with the top riser 13 and step down the 7-½" risers to riser 8, drawing the tread lines as you do so. The rail centerline can now be drawn to suit.

This method, making the stretch out within the narrowest possible space, is applicable to any layout whenever it is not practical to make the stretch out above the plan itself.

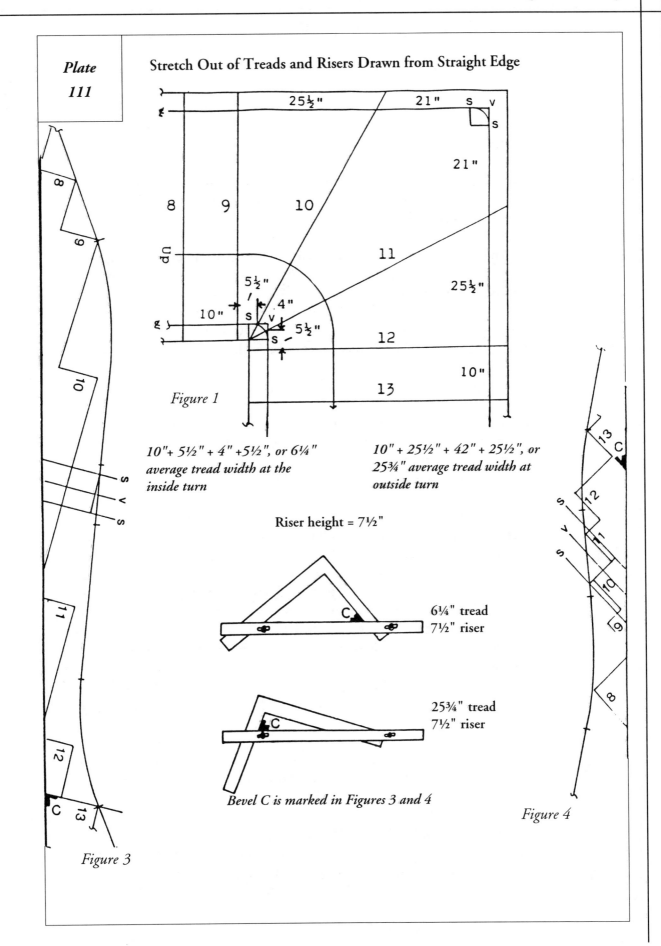

Plate 111

Stretch Out of Treads and Risers Drawn from Straight Edge

Figure 1

10" + 5½" + 4" + 5½", or 6¼"
average tread width at the
inside turn

10" + 25½" + 42" + 25½", or
25¾" average tread width at
outside turn

Riser height = 7½"

6¼" tread
7½" riser

25¾" tread
7½" riser

Bevel C is marked in Figures 3 and 4

Figure 3

Figure 4

Plate 112—Finding the Vertex Locations of the Level-to-Rake Handrail Sections at the Plan

Figure 1 is a plan of the handrail side of a curved stairway inclining from right to left with a starting volute section to a landing section of handrail at riser 12. Both are rake-to-level handrail sections.

Tangent lines are first drawn at the plan with the spring lines also being joint locations. Whenever possible in great radius stairs, spring line joints should try to be located at the risers if possible. In doing so, the normal rake rail height is established at these points. In the plan in **Plate 112**, spring lines are at risers 3, 7, and 11, as shown in **Figure 1**. Tangents GF, FE, ED, and DC are then drawn. Vertex points H and B have yet to be established.

From any convenient line as a level line, such as an extension of tangent GF, stretch out all plan tangents thus drawn, showing the riser intersecting points, vertices, and spring lines. Extend them into elevation as shown in **Figure 2**. Starting at any height along the vertical extension of riser 12, step down each successive riser to riser 2. Draw the pitch of the tangents through risers 3, 7, and 11, showing joints to be at C', E', and G'.

With the normal rake rail height of 2'-8" already established, the 3'-0" level rail height centers can be located at both the top and bottom as the line at which the vertices must be located for the rake-to-level sections. Therefore, H' at the top and B' at the bottom are vertex locations for their respective sections. H' and B' are then transferred to the plan in **Figure 1** as H and B. Level tangents HI and BA are then drawn. Granted, there will be an abrupt transition from level-to-rake at the volute section joints from A to C at the concave side of the rail section because of the desired 3'-0" level rail height and the steep pitch. This, however, can be overcome somewhat by raising the level rail height to say 3'-2". To accomplish this, simply raise the centerline along the plumb bevel at joint A by 2". This will necessitate additional block thickness, much like that shown in **Figure 102**.

A good rule to follow in making rake-to-level handrail sections is to be sure to make the sections long enough to offer greater latitude in order to make the most graceful transition ramp possible.

In any rake-to-level handrail section, whenever the height of the level volute or turn-out cap over the starting tread has not been predetermined, the level height to form the most graceful easing to the rake rail can be found by first making a stretch out ramp of the concave side of the rail, as in **Plate 105**.

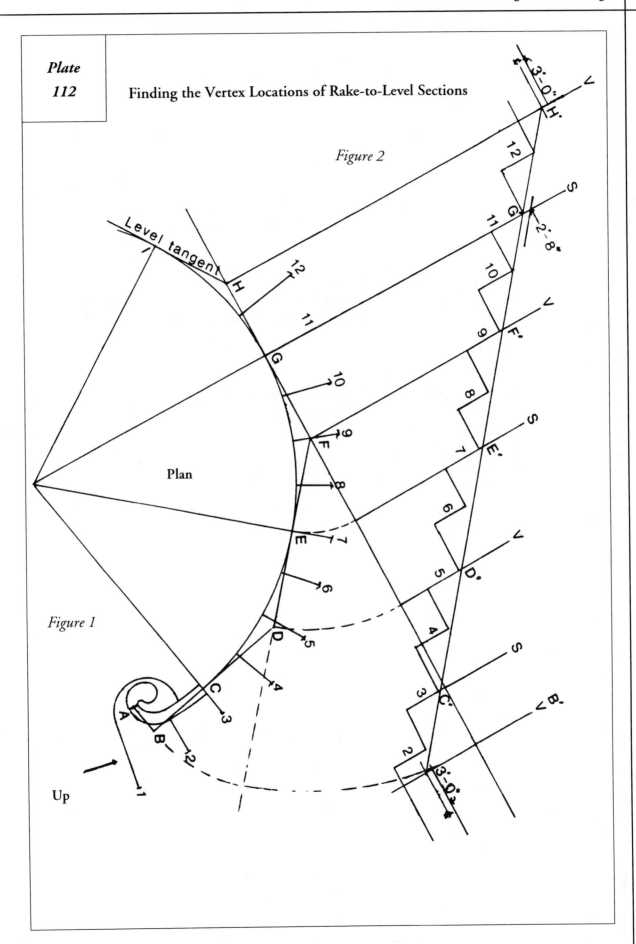

Plate
112

Finding the Vertex Locations of Rake-to-Level Sections

Figure 2

Level tangent

Plan

Figure 1

Up

Plate 113—Locating the Vertex for a Rake-to-Level Handrail Section in a Starting Volute

The handrail for a portion of a curved stairway is shown at the plan of **Figure 1** with tangents drawn between spring lines A and E. The vertex position of the rake and level tangents has not yet been established.

Stretch out the plan tangents along the line of tangents BC and CD. Extend the spring lines, vertices, and riser positions at the tangents into elevation in **Figure 2**. Step down the risers and draw the normal pitch of the centerline from risers 2 and 7. Let the normal rake rail height be 2'-8" over riser 5. The level height is to be 3'-0" above the starting tread, the centerline of which is line HI.

Now, to find the most desirable vertex position along HI. If tangent D'E' was to continue along the pitch of C'D', it would strike HI at J'. J', transferred to the plan at J, would require the short level tangent JK to the level cap. However, if tangent D'E' was raised above riser 2 approximately 2" to strike the centerline HI at, say F', then F', transferred to the plan at F, would make the level tangent FG. EF and FG would then be a much better plan tangent arrangement, allowing a slightly better transition from rake to level within a lower pitched turn.

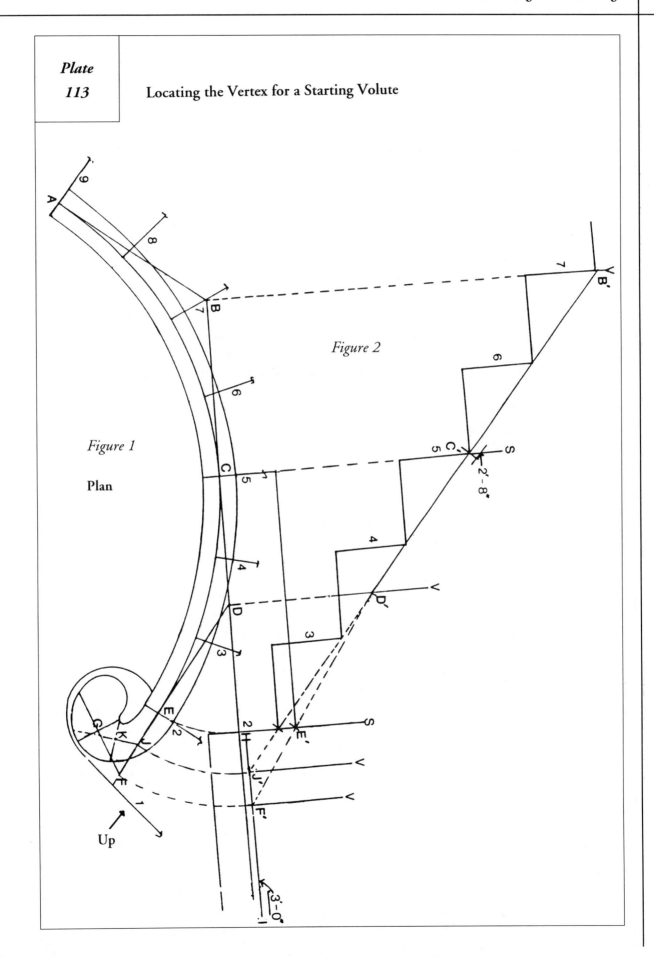

Plate 113

Locating the Vertex for a Starting Volute

Figure 2

Figure 1

Plan

Up

**Plate 114—A Quarter Circle Stair with Six Winder Treads
with Runs of Square Treads at Both Top and Bottom—Baluster Spacing
To Be Equal Throughout**

Assume in this quarter circle plan there are to be six winder treads within the spring lines of the turn with square treads below and above the spring lines. In order to have equal baluster spacing at both the winder and square treads without making the winder treads narrower than the building code may allow, it will be necessary to have three balusters per 10" square tread. Assume the distance between risers of the winder treads at the face of the mitered-face stringer to be no less than 6". Since the distance between 3 balusters of a square tread is approximately 6-5/8", let 6-5/8" be the distance between winder risers at the face of the stringer also. This would then make the radius approximately 25-5/16" to the face of the stringer and the balusters. All balusters can now be equally spaced.

In this plan, risers 6 and 12 are located at the spring lines. With the risers for the six winder treads spaced 6-5/8" apart at the stringer face, the walking line along the winders is approximately the same as that of a square tread 15" from the handrail center.

Figure 2 is a larger scale drawing of a winder tread at the stringer face showing the 6-5/8" dimension between risers.

Figure 3 shows the stretch out of the tangents and risers and their extension into elevation. The handrail is to be made up of one turned section approximately 6' in length. Although the normal pitch of the square treads is the same above and below the spring lines, the pitches do not align because of the narrower winder treads. In order to bring them into alignment without the use of easements beyond the spring lines, the adjoining straight rail can be cut on a bevel X as at **Plate 110,** and the transition easement made within the turn. In this manner the easing within the long section will be gradual. The block thickness must, however, be increased slightly over that indicated by the rectangular size of the rail at the bevel. See **Figure 2, Plate 68.**

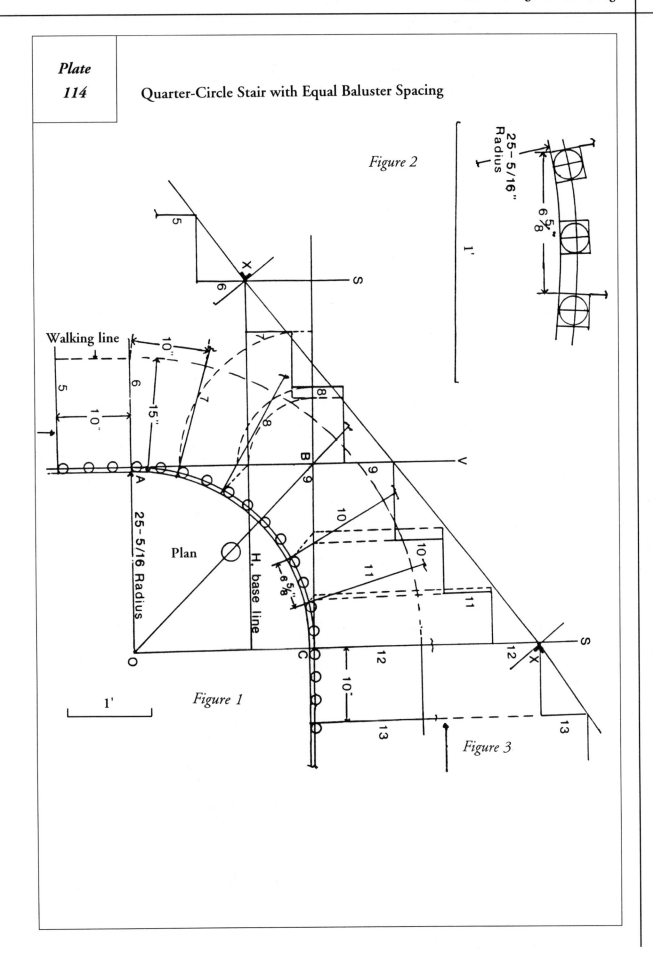

Plate 114

Quarter-Circle Stair with Equal Baluster Spacing

Figure 2

Figure 1

Figure 3

Plate *115*	Locating the Joints and Vertex Positions

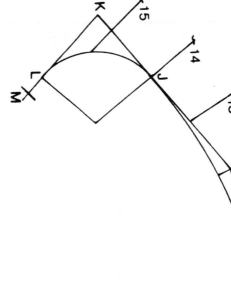

Plates 115 and 117 are the same centerline curves of a handrail for a circular staircase. In deciding just how many sections there should be in any circular stair where top and bottom rake-to-level sections are involved, the first step is to establish the rake joint locations of the rake-to-level sections.

In both **Plates, 115 and 117,** let rake joints be established at risers 2 and 14 for the rake-to-level sections. Also, in both plates, establish at least one spring line joint at a riser. In **Plate 115,** the joints are at risers 2, 6, 10, and 14, or D, F, H, and J. In **Plate 117,** the joints are at risers 2, 8, and 14, or D, F, and H. In **Plate 115,** there are three sections between the rake-to-level sections, whereas in **Plate 117** there are only two.

Plan

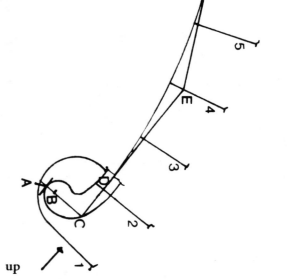

up

Plate 116

Locating the Joints and Vertex Positions

Plate 116 is the stretch out and elevation of the plan tangents and plan risers of **Plate 115**. Since joints are at risers 6 and 10, as shown in the plan of **Plate 115**, let the normal rail height be established as 2'-8". Draw the centerline pitch from risers 4 to 12 through risers 6 and 10.

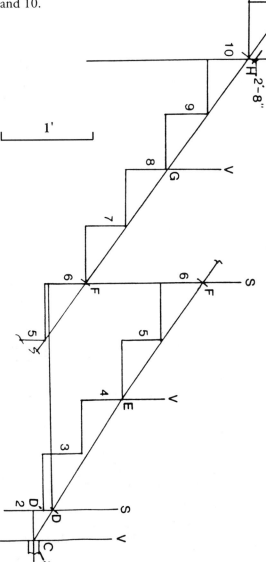

A 3'-0" level rail height is desirable over the first tread for the volute and balcony levels. Establish the 3'-0" level heights and draw the centerlines. From riser 4, draw the centerline pitch approximately 2" above riser 2 to intersect the volute centerline height, which will establish the vertex location C. Likewise, from the vertex I at riser 12, draw the centerline to intersect the level height centerline at K. Transfer JK' and CD' to **Plate 115** as JK and DC. At the turnout in **Plate 15**, CB equals CD, but CD is extended to A. At the top, in **Plate 115**, KL equals KJ. KL is extended to M so that the joint will not be at the spring line. Note that the length of tangents JK and DC in **Plate 115** determines the angle of the rake-to-level sections at the plan.

Plate 117

Locating the Joints
and Vertex Positions

In **Plate 117**, like **Plate 115**, tangents of sections between the rake-to-level sections are stretched out and extended into elevation in **Plate 118**. The normal rail height over riser 8, or F, is established and the centerline pitch drawn from the vertex at riser 5 to the vertex at riser 11, or from E to G. From E and G, the centerline pitches are drawn to meet the centerline heights at the volute and balcony levels at C and I. The level distances from C to riser 2, and from vertex I to riser 14, are transferred to **Plate 117** as DC and HI respectively, to determine the angle of the rake-to-level plan tangent sections. CA and IJ are the level tangents with IJ extended to K. Notice the slight difference in the pitch of the two tangents of a section by the dotted lines.

Whenever there is a slight plan curve, such as in these two plates, long rail sections should be used because of minimal cross-graining at the joints. Therefore, **Plate 117** would be a better choice of tangents than in **Plate 115**. However, regardless of which plan is chosen, the aesthetic appearance of the rail will be the same.

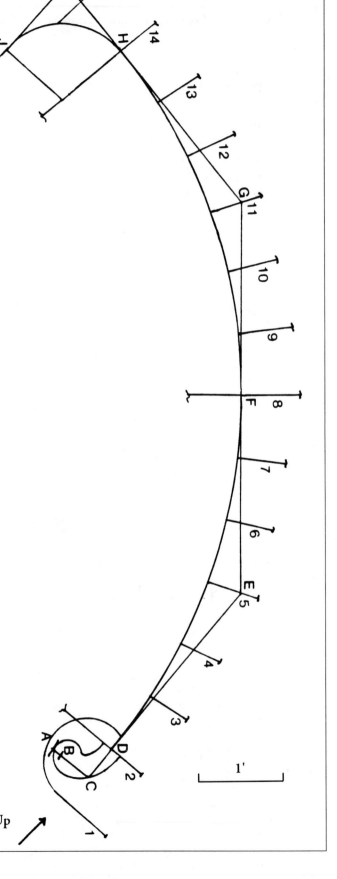

Up

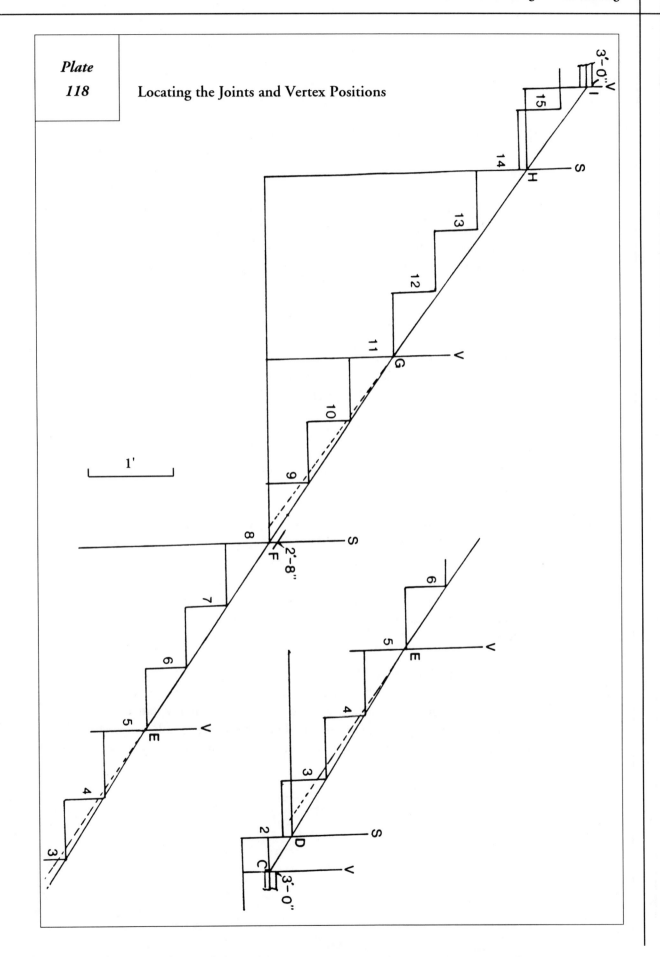

Plate 118

Locating the Joints and Vertex Positions

Plate 119—Short-Radius Stair with Five Incline-Turn Rail Sections

Figure 1 is a two-radius plan of a circle stair with five incline-turn handrail sections. In laying out the plan tangents, first locate spring lines D and H at the greater radius. Draw perpendicular lines to the radius at D and H and extend. Draw equal length tangents between D and H to make two equal rail sections, the tangents being DE, EF, FG, and GH. Establish an arbitrary joint location short of riser 15, say at J, riser 14. Perpendicular to radius PJ, draw tangent IJ and extend towards riser 15. Tangents IJ and IH are equal. The positions of the upper vertex K, and the lower vertex B, for these two rake-to-level sections, have yet to be established by the pitch of the tangents.

Figure 2 shows the stretch out elevation of the plan riser locations at the plan tangents and the vertical extensions of the spring lines and vertices. The elevation of all risers shows that pitched tangents of all sections can be in the same line with little difference in the height above the risers. Arbitrarily, let the normal rail height of, say 2'-8", be established over riser 6. 3'-0" level rail height centerlines are drawn over the first tread and at the balcony floor level. The intersection of the pitched and level centerlines establishes vertex positions B' and K'.

The level distances from riser 2 to vertex B', and from riser 15 to vertex K, are marked in **Figure 1** as B and K. Also in **Figure 1**, at the volute section, let the level cap joint be parallel to BC at A. AB will then be the level tangent. Establish the rake tangent joint arbitrarily at C for the tangent length CB. At the upper rake-to-level landing section, JK is the rake tangent and KL (equal to JK) is the level tangent. LMN is the angle of the tangents for a level turn section.

The learner should practice drawing scale plans of various types of circular stairways with continuous handrailings, experimenting with different joint locations of tangents in the same plan to see how little, if any, tangent positions affect the outcome of the rail. The exception is in rake-to-level sections where the location of the vertex affects the pitch of the rake tangent.

Plate 119

Short-Radius Stair with Five Rail Sections

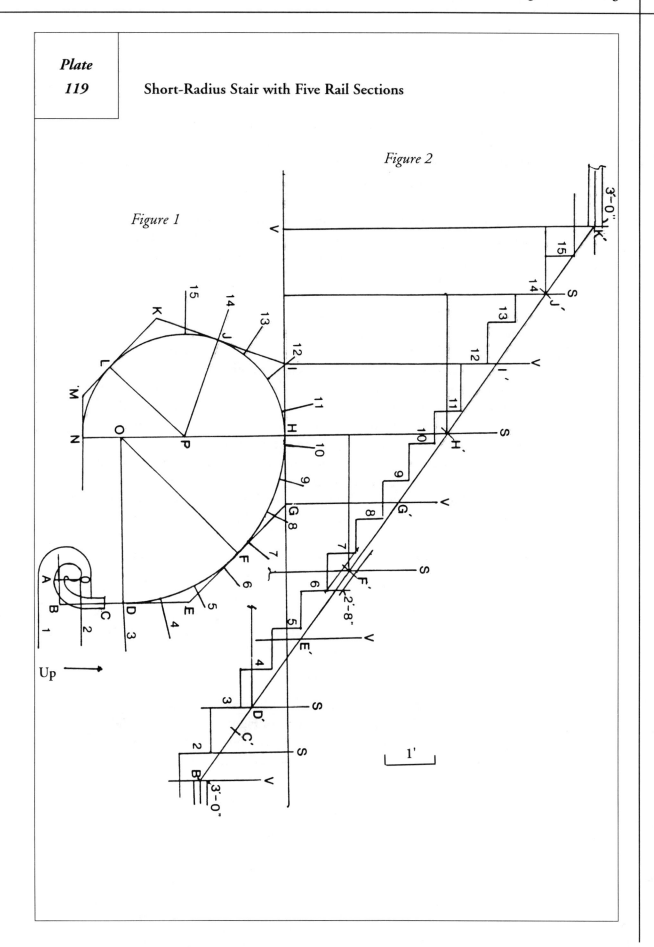

Figure 1

Figure 2

Plates 120 And 121—Tangent Pitches of the Plans of Plate 33

This is the first of a series of plan tangent placements and tangent pitches for plans of **Plates 33** through **40**.

Figure 1—The risers of this two winder plan are placed to satisfy equal baluster spacing throughout the stair, making the centerline radius 9-½". The spacing between risers along the tangent lines of the quarter circle centerline equals that of a square tread. Therefore, the pitch of the tangents will be equal as shown in the stretch out elevation.

Figure 2—In this 13"-radius quarter circle three winder stair plan, the winder risers are spaced to include two balusters shown at a square tread. Because of the plan riser locations, the normal pitches of both top and bottom square treads will not be along the same line, as shown at the stretch out elevation. This can be corrected by adjusting the rail to suit. **Plate 102** shows that by simply making an equal tangent layout, the joints at both A and B can be raised and lowered the distance x along the bevel line marked at the block joints. The correction can also be accomplished as in **Plate 114** by making an equally pitched tangent layout of the dotted tangent pitch and cutting the straight rail on the slight bevel at both A and B joints. With either method, the block should be slightly thicker than normally found by the bevels.

Plate 121 shows the 19" and 12" radius quarter-circle plans of **Plate 33** with four winder treads. **Figure 3** shows the winder risers at the spring lines, and **Figure 4** shows the winder risers 4" beyond the spring lines. The elevation of the risers and the pitch of the normal straight rail is drawn over the square treads in both figures. In **Figure 3** the normal centerline pitch over the square treads is also the pitch of the tangents, showing it to fall below the normal pitch of the upper square treads the distance x.

This misalignment is corrected similarly in **Figure 2** of **Plate 120** where the offset margin is split between both joints. In **Figure 4**, the straight rail is simply cut to the equal bevels at A and B, and the transition easement is made within the turn section. Normal block thickness must be slightly increased.

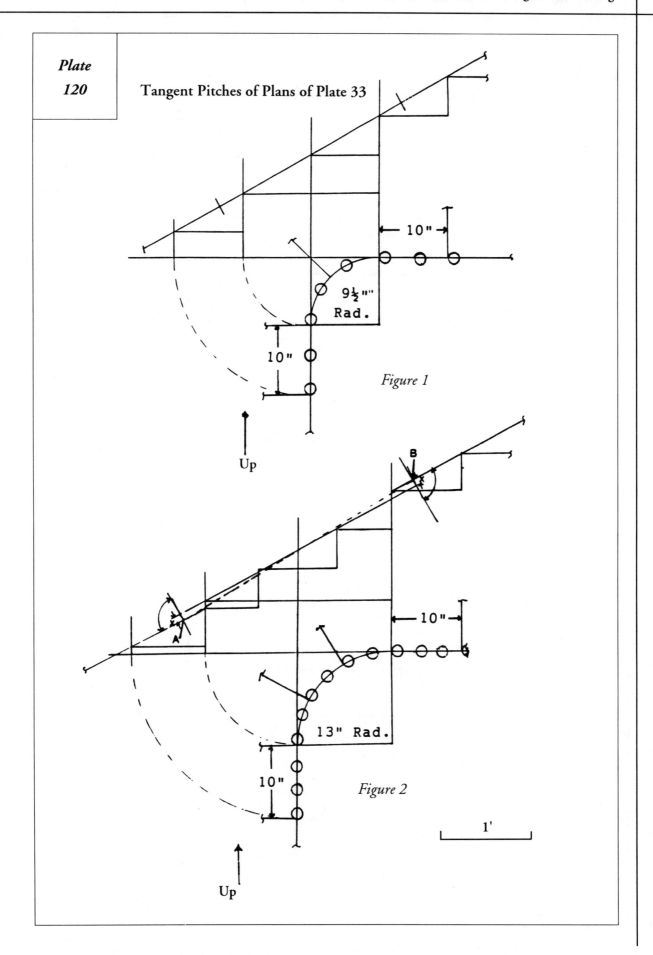

Plate
120

Tangent Pitches of Plans of Plate 33

10"

9½"' Rad.

10"

Figure 1

Up

B

10"

13" Rad.

A

10"

Figure 2

Up

1'

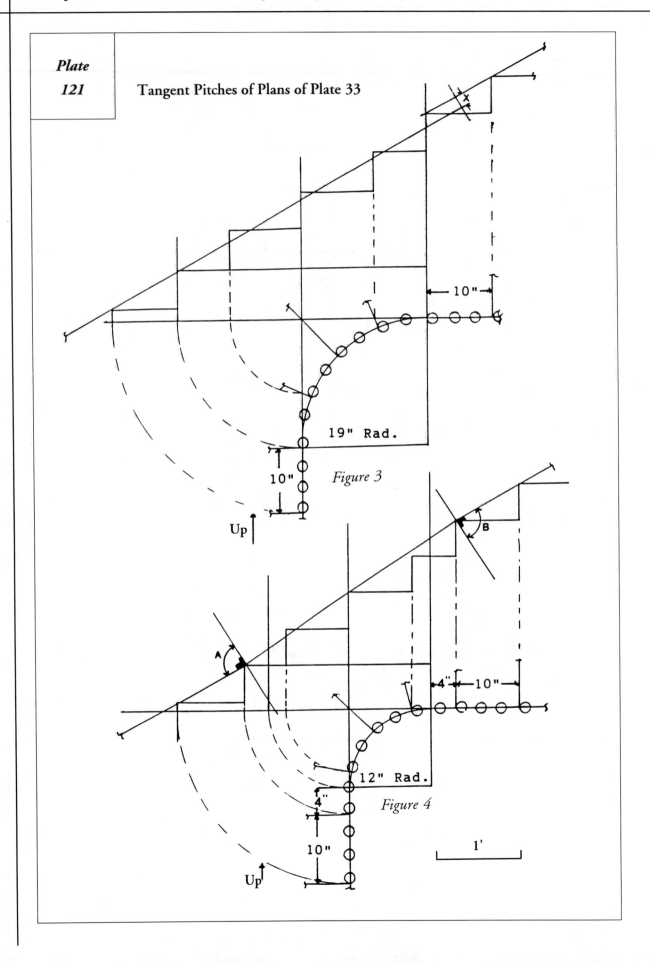

Plate 121

Tangent Pitches of Plans of Plate 33

10"

19" Rad.

10"

Figure 3

Up

A

B

4" 10"

12" Rad.

4"

10"

Figure 4

1'

Up

Setting up a 3"x4" white oak starting volute handrail in a floor vise to mark the top and bottom transition easement from rake to level.

Transition curves of the top of the volute from the rake to the level cap. (Cap is not shown here.) With the top of the volute completed, the rail is simply scribed to thickness to complete the "squaring" prior to shaping to the desired profile.

Plate 122—One Section Handrail for the Stair Plan of Plate 34

In **Figure 1**, the incline-turn handrail for this stair covers five winder-type treads. Although the rail can be made up of two sections, with one rake-to-rake section of plan tangents CQ and QP, and one rake-to-level turnout section of plan tangents PR and RG, I will demonstrate how the entire rail can be made up of one section only. R, P, and Q are then to be disregarded.

Extend the line of the plan rail centerline as the stretch out line of the plan tangents in **Figure 1**. At plan riser 8, mark up seven risers and draw the square treads in elevation between risers 6 and 8 in **Figure 2**. Draw the normal pitch of the rail centerline and rail thickness over risers 7 and 8. Let 2'-8" be the normal rake height over riser 8. The centerline of the level rail cap over the starting tread can then be drawn. If the centerline of the normal pitch of the rail were to continue down to strike the level centerline, establishing the vertex position at X to make the desired 3'-0" level rail height, the top of the rail would be excessively high over risers 2, 3, and 4. (See **Plate 106** showing plan risers drawn parallel to the ordinate direction). Therefore, to establish the vertex at a more practical location, draw the ordinate approximately 2" forward of riser 2 through the approximate center of the rail-cap to intersect the plan centerline of the straight rail at B. This is the level tangent and ordinate. Let BA equal tangent BC. Extend B to intersect the level centerline at B'. Draw the horizontal base line B'C'. Draw all elevated riser positions from the temporary base line BC. The true riser positions along BC are determined as in **Plate 106** by risers being drawn parallel to the ordinate from the rail centerline at the plan. From vertex B', draw the pitch of the tangent to intersect the normal centerline pitch of the square treads at, say riser 7. The pitch is shown to be fairly uniform over all risers. Since the pitched tangent and the normal centerline pitch over the square treads do not align, the straight rail is to be cut on a slight bevel at I.

It is therefore shown that the location of vertex B, along the level rail centerline, determines the height of the rail over the risers as well as the aesthetic appearance of the rail. The following plate shows the most desirable vertex location in a similar plan.

In **Plate 122**, square point A in **Figure 1** to E at the horizontal base line in **Figure 2**. To find the angle of the pitched and level tangents, make EF square to B'D, with B'F equal to BA in **Figure 1**. FB'D is the angle of the level and pitch tangents. B'H equals BG in **Figure 1**.

Now to find points through which the face mold curves may be drawn. From

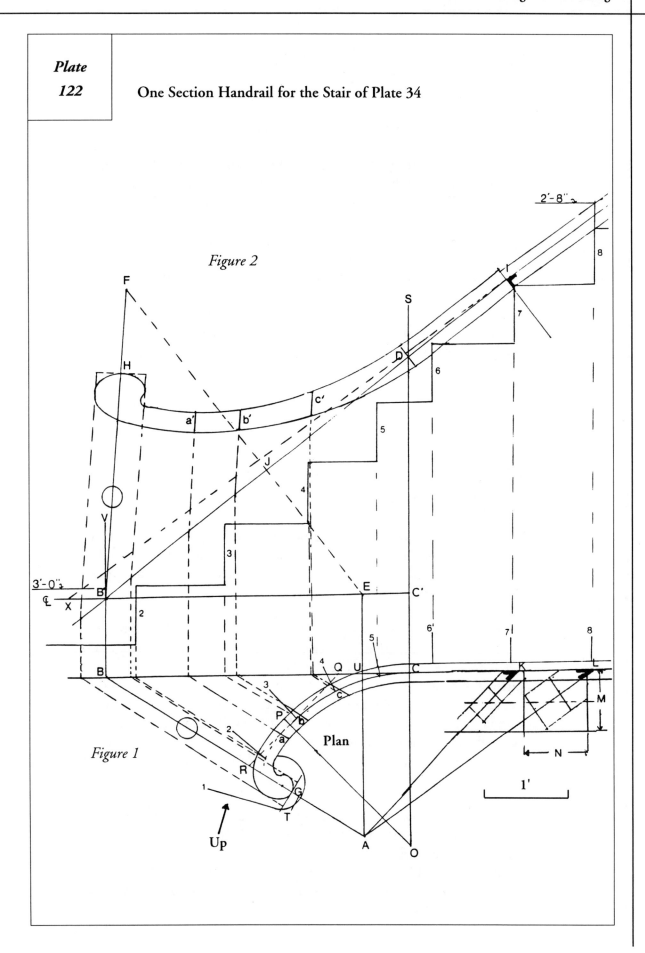

Plate 122

One Section Handrail for the Stair of Plate 34

Figure 2

Figure 1

Plan

Up

random positions <u>a</u>, <u>b</u>, and <u>c</u> at the plan in **Figure 1**, draw parallels (shown dotted) to the ordinate to intersect the rake tangent BC. They are extended vertically to strike the pitched tangent B'D in **Figure 2**, then drawn parallel to B'F the same lengths from the pitched tangent as they are from the plan tangents in **Figure 1**. The widths at <u>a</u>, <u>b</u>, and <u>c</u> are marked.

The bevels are found by referring to both plan and elevation. Let UK equal EJ for bevel UKA for the upper joint at I. The width of the rail at this bevel along the stretch out line is transferred to the face mold joint in **Figure 2** elevation so that the face mold curves can be completed. The bevel for the lower joint is found by making UL equal C'D. The width of the rail at the level cap is drawn parallel to bevel LA, making the block width N and the block thickness M.

Plate 123—Methods "A" and "B" Layouts of the Face Mold for the Turnout of Plate 35

The face mold shown above the pitched tangents is found by method "A", as in **Plate 122**. The face mold shown below the plan is found by method "B", as in **Plate 58**. The face mold in this plan is to include the entire quarter-circle plus the level turnout cap.

Let the centerline of the straight rail at the plan also be the line of the first tread as drawn in elevation. At riser 8 of the plan, step up seven risers to riser 8 and draw the tread line between risers 7 and 8. Now, draw the normal pitch of the rail over risers 7 and 8 with the centerline pitch drawn as shown. Let 2'-8" be the normal rake rail height at riser 8. The height of the level turnout rail-cap is to be 3'-0" over the first tread. Extend spring line OC vertically. Draw the centerline of the level rail-cap as B'C', which is also the horizontal base line, with B' the vertex of the level and rake tangents. Plumb B' to B.

Draw BA, as the level tangent and ordinate direction, forward of plan riser 2 to pass through the rail-cap as shown. BA equals BC. With the level tangent and the ordinate direction now established, the positions of the plan risers are drawn parallel from their intersecting points along the plan centerline to strike tangent BC, from where they are drawn into elevation as shown.

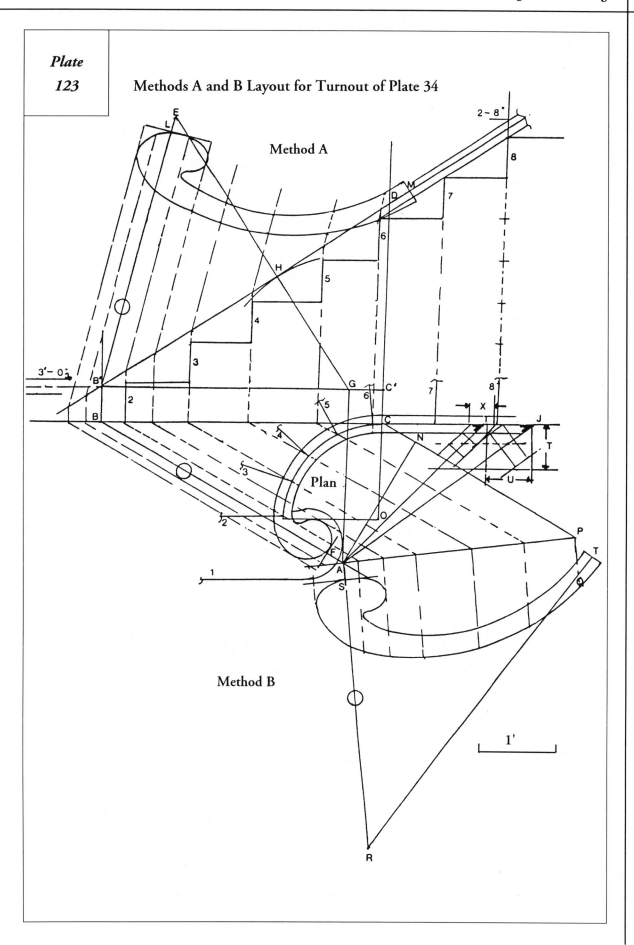

Plate
123

Methods A and B Layout for Turnout of Plate 34

Method A

Plan

Method B

1'

Plate 124—The Rail of Plate 35 Made Up of Three Incline-Turn Sections

While **Plate 123** shows the starting turnout and adjoining quarter turn of the plan in **Plate 35** being made as one section, **Plate 124** shows the same turnout and quarter-turn made of two sections. At the top of the same stair (not shown in **Plate 123**) is a quarter-turn section to butt to a balcony post.

Figure 1 is the stair plan of **Plate 35**. In establishing the plan tangents, let the level-to-rake turnout section be of workable size by locating a joint, say at C, to make both tangents of the obtuse rake-to-level section equal. The tangents of the turnout section can be at right angles, with AB, the level tangent, slightly longer than the rake tangent BC. AB is forward of riser 2 so that the vertex is <u>over the starting tread</u>. CD and DE are equal-length plan tangents of the obtuse rake section.

The plan tangents of the upper quarter-turn section are HI and IJ. The upper tangent of this section will include enough straight rail in the face mold beyond J so as to butt into the balcony post at riser 15

Plan risers, spring lines, and vertices for all sections are drawn to intersect line XY and then extended into elevation in **Figures 2** and **3.** The normal pitch over the square treads is drawn from risers 6 to 9 and extended to intersect the vertex of the obtuse section at D' in **Figure 3** elevation. In **Figure 2** elevation, the pitches of both upper quarter circle tangents are the same. A slight ramp, E'G', is required between the upper quarter-turn and the lower rake-to-rake obtuse turn.

In **Figure 3**, the upper tangent pitch of the obtuse rake-to-rake section is E'D'. From D' draw the lower tangent pitch D'C' so that the top of the level cap turnout is 3'-0" above the starting tread at vertex B'. The 2'-8" normal rake rail height is at riser 9 of **Figure 3** and riser 14 of **Figure 2**. The face mold for the turnout section is found by drawing B"C" arbitrarily above B'C' and drawing parallels the same lengths as at the plan in **Figure 1**. The bevel at C' of **Figure 3** is drawn along XY at N, showing the off-center application as in **Plate 123**. QR is the block width and PQ the block thickness. Joint C does not require a bevel.

The horizontal base line for the rake-to-level quarter-turn section is B'K. C'L is the base line for the obtuse turn. H'M is the base line for the upper quarter-turn section in **Figure 2**.

The face mold of the turnout section is found as in the two previous plates,

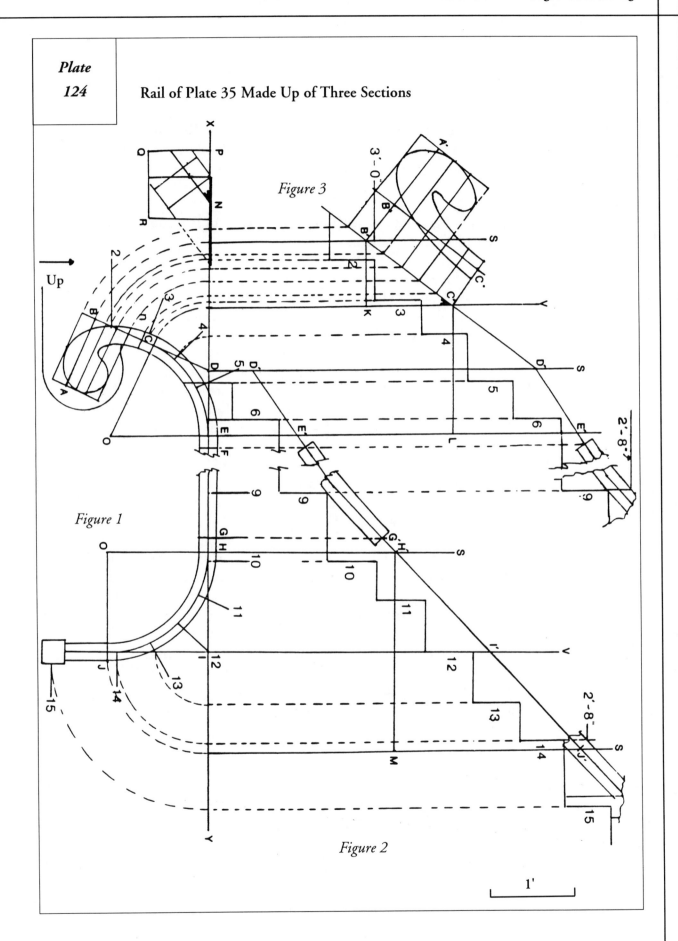

**Plate
124**

Rail of Plate 35 Made Up of Three Sections

Figure 3

Up

Figure 1

Figure 2

1'

with B'C' transferred to the pitched tangent as B" and C". Refer to the chart in **Plate 72** for the face mold layouts for the rake-to-rake obtuse section and the equally pitched upper quarter-turn section.

Plate 125—The Plan and Pitched Tangents of Plan 4 of Plate 36, the Rail To Be Made in Three Sections

Figure 1 is the same Plan 4 shown in **Plate 36**. Tangents for the handrail are drawn to make a long flat obtuse curve section, a rake-to-level acute section, and a separate level turnout rail cap. Risers 9, 10, and beyond are between square treads. The rail centerline of the square treads is extended as AB, with C a spring line of the flat curve section. From line AB, DE is made equal to tangent DC, with spring line E falling exactly at riser 3. ED and DC are, therefore, the tangents for the upper obtuse section. Extend DE to F. Let G be the joint of the level cap and level-to-rake section. Draw GH perpendicular to the joint width. HI is then the level tangent and equal to HE.

In **Figure 2**, the shaded right triangle shows the pitch to be determined by the average tread width along the stretched out risers along the tangents in **Figure 1** and the normal riser height. The stretch out elevation of the treads and risers can then be drawn along the straight line XY.

With the treads and risers thus drawn, the pitch of the tangents of the long obtuse section is shown to be the same as the normal centerline pitch of the rail from risers 9 and above. The normal rake rail height is 2'-8", and the level rail height at the turnout cap is to be 3'-0".

With MH in **Figure 2** equal to EH in **Figure 1**, the continued pitch of tangent DE shows the vertex H to be well below the desired centerline height at J. This difference in height can be overcome by simply raising the plumb line joint of the applied bevel the distance HJ.

The face mold will not be drawn in this plate. However, it can be drawn by either method "A" or "B" in **Figure 1**.

To find the bevel for the level joint at G in **Figure 1**, draw NI perpendicular to tangent HE. Let NP equal height ME in **Figure 2** and connect bevel line PI for the bevel NPI. For the rake tangent bevel at joint E, let MN' in **Figure 2** equal EN in **Figure 1**. Let NO' in **Figure 1** equal arc N'O in **Figure 2**. Connect O'I for the required bevel NO'I.

From both sides of bevel line PI in **Figure 1**, draw half the plan rail

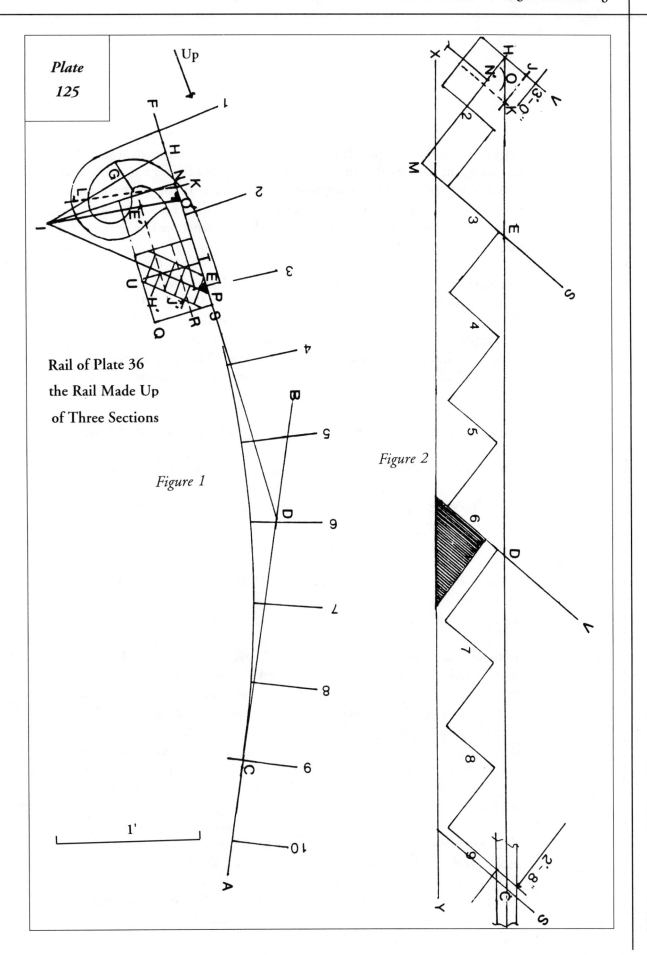

Plate
125

Up

Rail of Plate 36
the Rail Made Up
of Three Sections

Figure 1

Figure 2

1'

width. Draw the rectangular rail size at the top first, with the centerline as J'. Let J'H' equal JH in **Figure 2**. From center H' draw the rectangular size of the rail again, showing that the normal block thickness for centerlines H' and E' would be QR. However, since center H' is raised to J', the block thickness must now be QS. Although J' is the raised center at the level joint G, E' remains as the rail center at the rake joint. TU represents the line of the tangents at the block surfaces.

In **Figure 2,** if K was to be the vertex, there would be no need to raise the joint to J. With position K transferred to the plan in **Figure 1** the same distance from riser 2, the level tangent (shown dotted) would pass through the approximate center of the rail-cap as KL, equal to KE. The entire section would then include the rail-cap as shown in **Plate 124.** However, the purpose of this rail is to show the rail-cap as a separate level segment.

Plate 126—The Plan and Pitched Tangents of Plans 5 and 5-A of Plate 36

Plans 5 and 5-A, showing a platform between runs of square treads, are the same as those of **Plate 36**. The pitch of the tangents are to be shown in this plate. The risers are spaced so that balusters are equally spaced throughout the stairs at each plan.

In Plan 5, draw the quarter-turn tangents AB, BC, CD, and DE. Draw the layout line XY through tangents BC and CD. Stretch out all plan risers and spring lines A and E along XY and extend into elevation as shown. The normal rail pitch is at square tread risers 1 and 2 and at 3, 4, and 5.

With the rake rail height established as 2'-8" over both risers 1 and 5, the level rail height at vertex B' for the lower rake-to-level section is found to be the desired 3'-0" at joint C'. Whereas, the rake-to-level upper section shows that vertex D' will be considerably below vertex B'. Consequently, the level joint C" must be raised plumb to meet the desired height at joint C' of the lower section. This can be accomplished by making the block for the upper section thicker than that required for the lower section. The difference in height can be adjusted along the plumbline bevel that is to be applied to the level joint of this upper section.

The bevel shown at E' is the bevel for the level joint of both sections. The rake joint does not require a bevel. Draw the rectangular size of the rail with F the center. The block thickness H will be that of the lower section only. For the block thickness for the upper section, make FG, along the applied

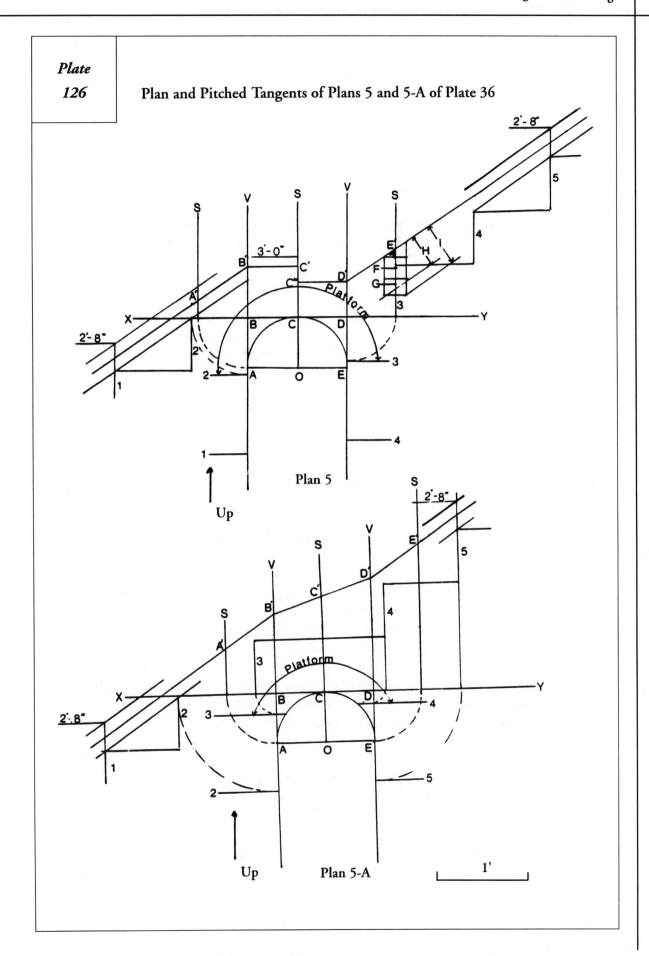

Plate 126

Plan and Pitched Tangents of Plans 5 and 5-A of Plate 36

Plan 5

Up

Plan 5-A

Up

1'

bevel, equal to C'C". Draw half the rail thickness below G, showing the minimum block thickness to be I. The block width must include the rail sizes shown at both F and G. The margin from center G to the bottom of the block is the same to center E' at the rake joint.

In **Plan 5-A,** extend the risers and spring lines into elevation as in **Plan 5** from XY. Extend the normal centerline pitch of the square treads to intersect the vertex lines at B' and D'. Connect B'D', making C' the joint of both lower and upper tangents of both sections.

Since the corresponding tangents are the same at both sections, only one face mold is required for **Plan 5-A.** One section is simply the reverse of the other. The bevels found for one section also apply to the other.

Plate 127—The Plan and Pitched Tangents of Plans 6 and 7 of Plate 37

Plan 6 shows the centerline of a rail for a half-circle-stair to be made up of two quarter-turn handrail sections with equal plan tangents AB, BC, CD, and DE.

The stretch out of the plan risers and tangents is made along a straight line as shown. The shaded triangle is the pitch of the normal riser height and the average tread width between risers along the plan tangents. At the stretch out, the normal centerline pitch is drawn over the lower and upper square tread risers from 1 to 2 and from 9 and above. Since riser 5 is at the spring line, the normal centerline height is established here also. The tangent pitches are then drawn to meet the normal pitches. This is best accomplished by pitching tangents BC and CD to intersect the vertex lines as shown, and then pitching tangents AB and DE equally so as to make slight easement adjustments beyond the spring lines to meet the normal pitch. By making both sections with identically pitched tangents, only one layout to find the face mold is required.

In **Plan 7,** the plan tangents for this quarter circle rail shows the rail to be made up of a single quarter-turn section with tangents AB and BC and slight easements beyond the spring lines.

An option, in **Plan 7,** which eliminates beyond spring line easements, is to make a bevel cut at the straight rail as shown in **Plate 114,** and then increase the block thickness in order to make the easements within the turn section.

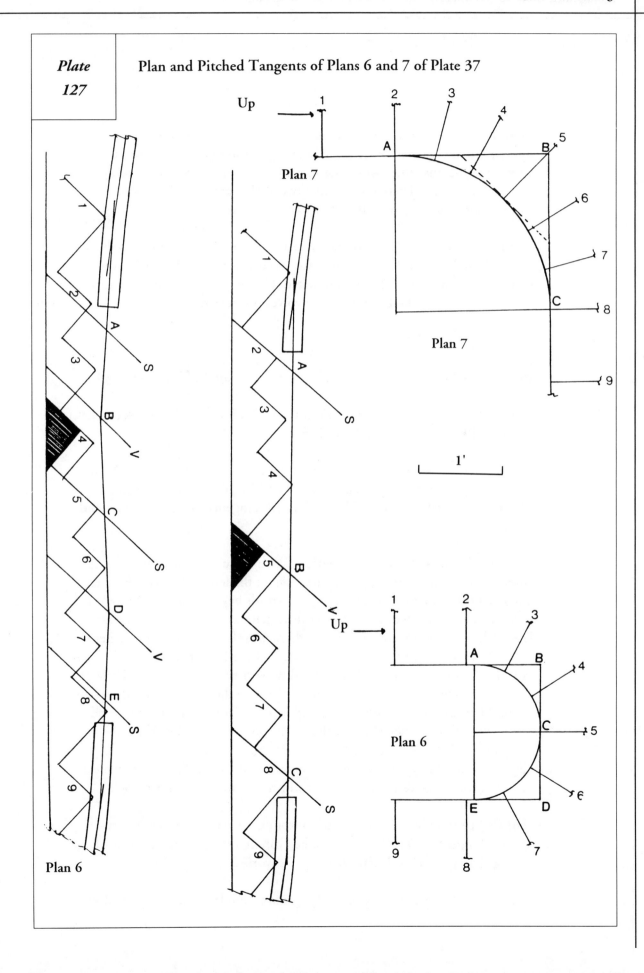

Plate 127

Plan and Pitched Tangents of Plans 6 and 7 of Plate 37

Plan 7

Plan 7

1'

Plan 6

Plan 6

However, because of the great parallel distance between both lower and upper normally pitched centerlines, beveling the straight rail joints is not a wise choice because there would necessarily be excessive "S" curve easing between the joints. Therefore, making two slight easements beyond the spring lines would be the better choice.

The handrail in **Plan 7** can also be made up of two obtuse sections as shown by the dotted tangent lines. But the two easements required by using the quarter-turn layout will not be eliminated. Consequently, the quarter-turn section is the wiser choice because it involves one less joint.

Plate 128—The Plan and Pitched Tangents of Plan 8 of Plate 38

In this plan, the procedure to establish the vertex for the starting rake-to-level section is like that shown in **Plates 113, 116,** and **118.**

Assume that the level rail-cap joint at rail center A is established first. Next, draw the upper quarter turn tangents GF and FE. Extend FE to make an obtuse rake-to-rake section with equal length tangents ED and DC. Extend DC.

The stretch out of the tangents, vertices, spring-line positions, and riser intersections at the plan tangents are shown at the left of the plan. All are extended into elevation. There are square treads beyond riser 11. Since it is obvious that the centerline pitch over the square tread risers is not the same as the pitch over the stretched out risers, there must be a slight over-easement beyond the upper spring line G, as shown. Let the 2'-8" rake rail height be at riser 11. Draw the 3'-0" level rail centerline over the starting tread. Draw the pitch of tangents GF, FE, and ED approximately the same height over risers 7 and 11 to intersect the normal centerline of the upper easement beyond the spring line. From vertex D, draw the extension of tangent DC the same height over riser 2 as at riser 11 to intersect the level centerline at B. The level distance from B to riser 2 is then transferred to the plan. BA', at the plan, is drawn through the center of the joint width at A to equal tangent BC.

The slight easing in the obtuse section from C to E is not noticeable, although tangents CD and DE are pitched differently. The section is just large enough so that a smooth transition can be worked.

Whenever there is a starting or landing rake-to-level section in a curved stair

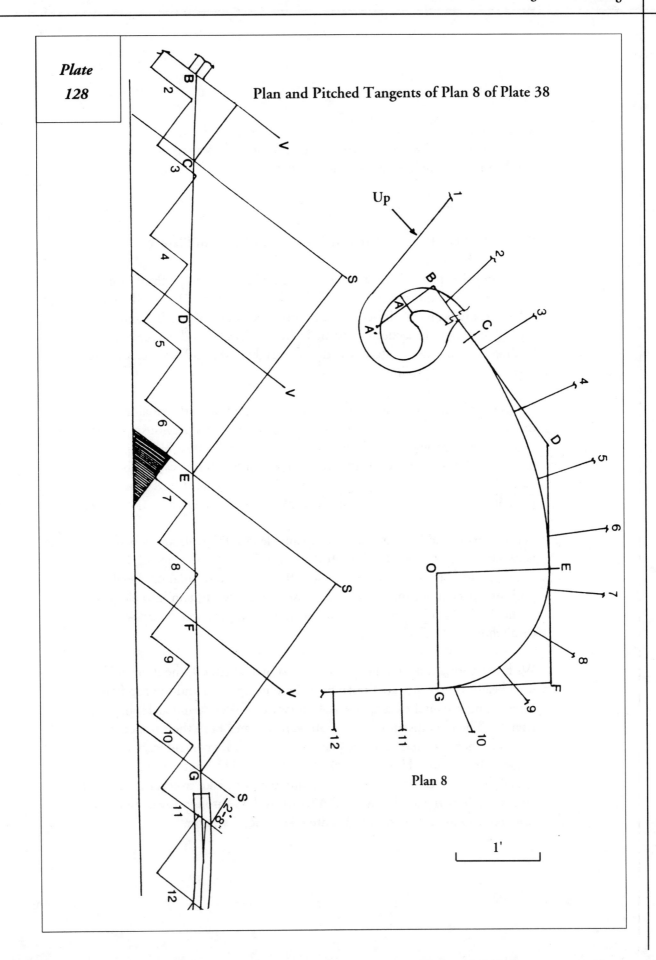

Plan and Pitched Tangents of Plan 8 of Plate 38

Plate
128

Up

Plan 8

1'

where there are a number of vertices where pitches can be altered, the vertex of the rake-to-level section can often be positioned as desired prior to drawing the pitch of the tangents without any effect.

Plate 129—The Plan and Pitched Tangents of Plan 9 of Plate 39

This is another plan where tangents must first be drawn to pitch in order to locate vertex positions of starting and landing rake-to-level handrail sections. It has already been established that the vertex positions of starting and landing rake-to-level rail sections must be in front of riser 2 and beyond the landing riser in order to attain the desired heights over both the starting tread and the top floor landing.

Let us start by first locating a joint at G, the spring line of the small radius of the upper quarter-turn. Draw extended tangents at G. As previously stated, rake rail joints should be positioned at riser lines, and if possible, as points of normal rail height. However, this does not mean that joints cannot be located at any other point along the curve. In this respect, I have placed a joint at C in order to make the starting rake-to-level section a workable size.

With joints C and G established, four equal length tangents are now drawn between the joints, such as CD, DE, EF, and FG. Note that spring-line joints C and E are almost at risers 4 and 9, or close to normal rail height position. Stretch out the tangents and risers and extend them into elevation as shown. It is readily seen that equally pitched tangents can be drawn throughout.

With the normal-rake rail height of 2'-8" shown at riser 9, the desired 3'-0" level height centerlines are drawn over the starting tread and the top floor, intersecting the pitched tangents at B, forward of riser 2, and at H, beyond riser 15. The level distances from both vertex B and H to their respective risers are then transferred to the plan at B' and H'. By coincidence, the tangent drawn from H' to the centerline curve strikes exactly along the straight level rail centerline, to make a quarter-turn, rake-to-level section. At vertex B', draw the level tangent B'A' equal to B'C through the center of the level rail-cap joint A, making an acute rake-to-level section.

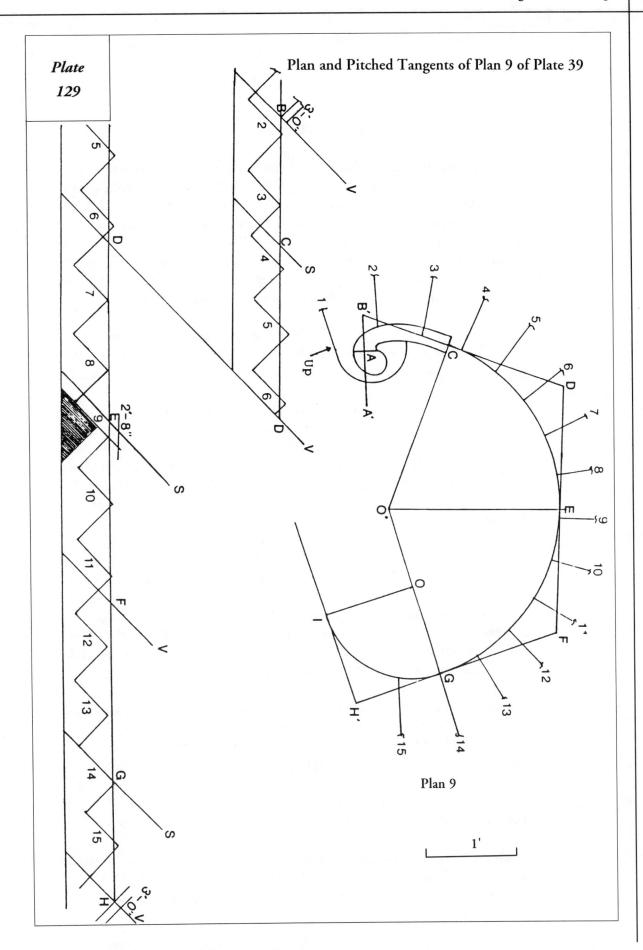

Plate
129

Plan and Pitched Tangents of Plan 9 of Plate 39

Plan 9

1'

Plate 130—The Elliptical Plan and Pitched Tangents for Plan 10 of Plate 40

The location at the plan for joints of an elliptical incline-turn handrail section shows that one plan tangent must be shorter than the other. However, in this method for finding the face mold and bevels for such a section, both plan tangents must first be shown equal length, and then the shorter tangent is reduced to its proper length in the face mold. This condition has already been explained in **Plate 88**. There will be no deviation in the method of finding the bevel(s) for any layout, whether the plan tangents are equal as in radius-drawn plans, or one tangent is shorter than the other as in elliptical plans.

However, in an elliptical plan at the option of the railmaker, the layout can also be made without having to lengthen the short plan tangent. The bevels, in this case, are found at the angle of the tangents, as shown in **Plates 89** and **90**.

Adhering to the method shown in **Plate 88,** let us proceed to draw the tangents for this elliptical plan. For the tangents regarding the outside curve, start by placing a joint at the minor axis at P. Draw the perpendicular OQ to the minor axis at P. Draw perpendiculars from N and R' to respective radius lines XN and X'R' as MO and QT. Let QR equal QP and ON' equal OP. Draw MK through L and TW through U. With the wall corner fixed, draw tangent WY perpendicular to the wall line.

The next step is finding the vertex positions for the starting volute section. To begin, the most desirable position for the joint of the level rail-cap is at J in order to allow more curved rail to make the gradual transition from the level joint to the rake joint at L. Perpendicular to joint J, draw the level tangent J'K to intersect the extension of tangent LM. With K positioned prior to making the stretch out elevation, let us see if the position has any adverse affect on the pitch of tangents ML and LK in the elevation layout.

Figures 1 and **2** are the stretch out elevations of the tangents and risers at the wide side of the stair showing the positions of the actual joints L, N, P, R', and U: the equal length tangent positions N' and R' for finding the face mold and the vertices. Let the normal rail height centerline be over risers 8 and 9 in **Figure 2**, which will determine the level centerline heights over the starting tread and at the top floor. Let the straight centerline pitch over risers 8 and 9 extend to meet the level centerline at vertex W and intersect the vertex at O.

It is seen that if this same pitch continues, it would be excessively high over the lower risers. Therefore, from vertex O, draw pitch OM approximately the same height over risers 6, 5, and 4, similar to the pitch over risers 8 and 9, to inter-

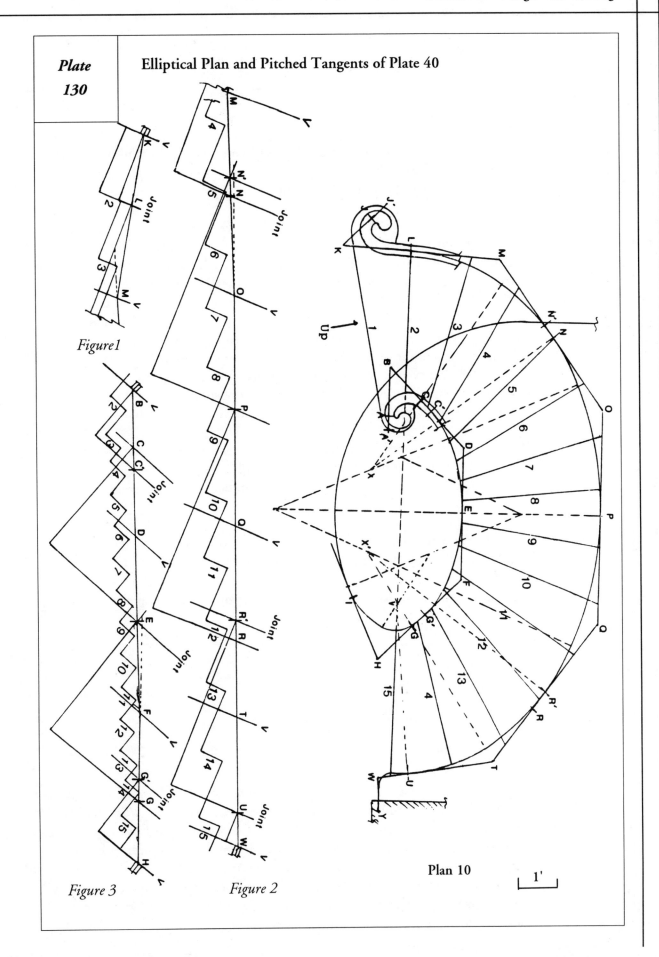

Plate 130

Elliptical Plan and Pitched Tangents of Plate 40

Figure 1

Figure 3

Figure 2

Plan 10

1'

sect the vertex at M. In **Figure 1**, draw the pitch from M to the already established vertex position K to meet the level centerline height. This will make the rail slightly higher over riser 2. Having established the plan vertex at K, it is found that the location does not have a negative affect on the outcome of the rail.

To draw the tangents for the inside curve, at minor axis point E, draw an extended perpendicular, as DF. As aforementioned, rake-to-level sections should be of substantial size in order to properly work the easing from the level to the rake joint. Therefore, random joints are located at C' and G', through which extended tangents to the curve are drawn, such as DC and FG. Tangent DC equals DE, and tangent FG equals EF for layout purposes. Like the rail-cap of the outside rail, mark a random joint location, say at A. To locate the vertex at B draw the perpendicular from A to intersect the extension of DC'. Let BA' equal BC'. The vertex position H has yet to be found. The stretch out elevation must now be drawn.

Figure 3 shows the stretch out elevation for the inside curve. All lettering corresponds to those at the plan. The normal rail height will be over risers 8 and 9. The centerline of the rail height at the top floor level can then be drawn.

Connect vertex B to vertex F at the proper height over risers 8 and 9. From F, draw tangent pitches over riser 14 to determine vertex position H at the floor level. Vertex location H is then transferred to the plan, with GH at the plan equal to the level distance from G to H at **Figure 3**. Level tangent HI at the plan happens to be the same length as HG'.

Residential stair at Lake Tahoe, California. Custom-made, self-supporting stair with 2-½" x 12" laminated Douglas Fir stringers veneered with 3/16" thick red oak. 3/8" sawn brackets at both sides of exposed stringers are mitered to a finished piece across the stringer riser. The 2-¼" thick red oak treads have a mitered return piece across the ends to receive the brackets. The curved and inclined 2" x 3" red oak handrail at both sides of the stair is laminated. Tangent-made starting turn-out and landing sections are connected to the laminated rail. 1-½" square top and bottom red oak balusters are alternating twisted and turned types.

Plate 131—Combined Use of Both Tangent and Laminated Handrail Sections

It is common practice today, mainly because of the cost-saving factor, for an architect, designer, or stairbuilder to design a stair using stock handrail parts and laminated handrails. Consequently, the design is restricted to the use of these parts. Although satisfactory results may be attained, very often the combined use of tangent-made and laminated handrail sections will present a more distinctive and aesthetically pleasing stairway.

In **Plate 131, Figure 1** shows a curved stair plan of the handrail side of a stair with 15 risers, starting with a volute and landing at the top of the stair with a rake-to-level continuous section. Both the volute and the landing section are to be tangent made and connected to a laminated handrail section. The starting volute consists of a level rail-cap and a separate level-to-rake turn and easing section. The combination of the volute, the laminated section, and the rake-to-level turn at the top offers the most graceful and eye-appealing handrail possible for this stair. See **Figure 5.** If stock parts were to be used with the laminated rail, the starting volute would consist of an abrupt straight-sided easing to the normal rail pitch (**Figures 6** and **7**), instead of a gradual easing through the turn from the level cap joint, as would be the case with the tangent-made section. Also, with stock parts, the landing turn would consist of a ramp section forming a gooseneck mitered to a level quarter-turn. Otherwise the laminated rail itself might be terminated into a gallery post.

In **Figure 1,** the procedure for making the volute and rake-to-level sections to join the laminated handrail is as follows: To make the volute section, draw tangent AB perpendicular to the level cap joint, say at A. Mark a rake joint, say at C. Draw a tangent to the curve through C, perpendicular to the radius line OC, if the radius can be drawn. Otherwise, continue the centerline curve towards riser 1, mark off equal segments on each side of the curve from C, which we will assume to be at risers 2 and 3, connect points 2 and 3 as a straight line, and draw a parallel to line 2-3 through point C. The parallel just drawn will be tangent to the curve at C. Draw CB to intersect the level tangent BA at B. Let BA' equal BC. Also in **Figure 1,** to find the vertex location E of the top rake-to-level landing section, it will be necessary to first make the stretch out in **Figure 2** of the plan centerline of **Figure 1,** showing all treads and risers in elevation. Assume that the normal rail height over riser 13 is to be 3'-0", and the top floor height is to be 3'-6" as shown. The centerline pitch of the rail intersects the level centerline at E as the vertex. DG will then be the distance to mark from D in **Figure 1** as DE. O'D or

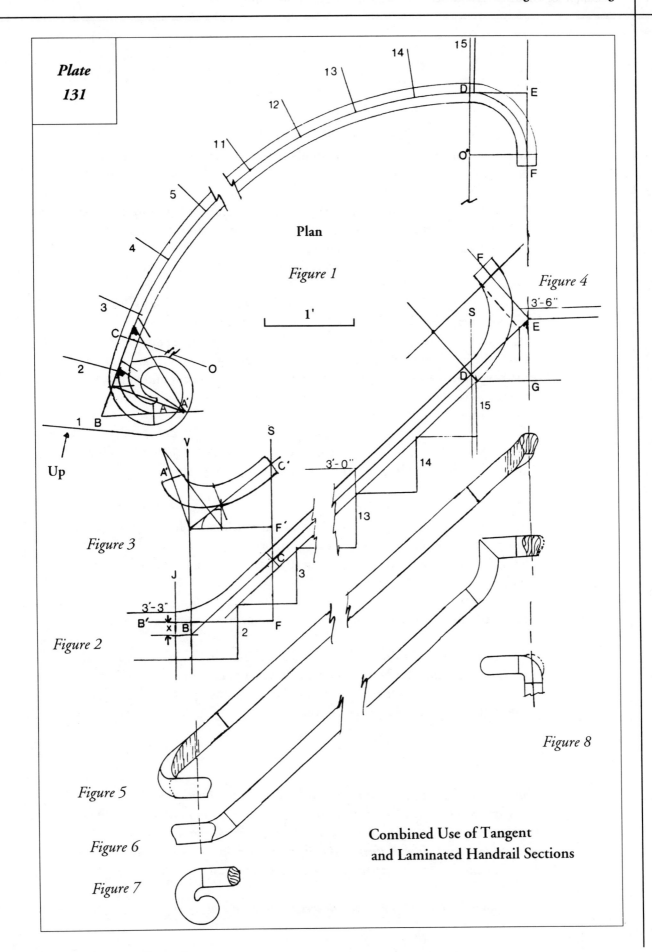

Plate 131

Plan

Figure 1

1'

Up

Figure 3

Figure 2

Figure 4

3'-6"

3'-0"

3'-3"

Figure 5

Figure 6

Figure 7

Figure 8

Combined Use of Tangent
and Laminated Handrail Sections

FO' is the radius of the quarter-turn rake-to-level section. In **Figure 4**, the face mold for the rake-to-level handrail section is drawn by the trammel method shown in **Plate 2**.

At the starting volute in **Figure 2** elevation, the vertical line J represents the stretch out centerline position of joint A in **Figure 1**. From the top of the rail at joint C draw a ramped easement to the joint line J as shown. If the rail height is desired to be the same as at the top of the stair, then the easement can be more gradual than is shown. In this instance, the rail height is shown to be 3'-3". Let FB equal CB in **Figure 1**. Draw the rake tangent CB to intersect the vertex line B. The face mold for the section is drawn in the elevation of **Figure 3**. C'F' equals CF in **Figure 2**. See **Plates 86** and **88**. Once the bevels are found and applied, the center of joint A is raised along the bevel line the distance marked x. Joints between tangent and laminated rail sections should always be assembled and dressed at the sides prior to working the top and bottom.

Figure 5 is a side view perspective of the tangent-made level-to-rake volute and rake-to-level landing sections joining the laminated curved railing.

Figure 6 shows an alternative side-view perspective if stock volute and gooseneck rail parts are joined to the laminated section. The gooseneck is mitered to either a stock level quarter turn or rail-cap rail part at the balcony level. A balcony post can be used beneath the rail for lateral strength regardless of which method is used.

Figures 7 and 8 are plan views of the stock rail parts.

The conclusion to be drawn from this plate is that although stock handrail parts can be satisfactorily used with a laminated rail section, very often the use of the tangent-made rail sections, rather than stock parts, will improve the aesthetic appearance of the entire stair.

Continuous 2"x3" Douglas Fir handrail around a platform turn of two straight runs of a carpenter-built staircase. This continuous handrail method has a more aesthetic appearance than an otherwise lower run gooseneck ramp mitered to a level quarter turn at the platform and an upper run easement joined to another level quarter turn at the platform.

Continuous hardwood handrail features the wood grain.

Plates 132 through 143—Twelve Non-Circular Stair Plans with Continuous Handrailings through Right, Obtuse, and Acute Angle Turns

The following nine plates of stair plans with continuous molded handrails are drawn so that balusters can be equally spaced throughout the entire stair. The remaining three plates are plans of full round handrails around a wall corner.

Plates 132, 133, and 134 show segments of stair plans with either the starting or landing risers correctly numbered in the entire stair. The remaining plans, **135** through **143**, are also segments of stair plans with intermediate risers marked as 1, 2, 3, etc., for the purpose of identifying riser positions only. For example, the riser marked number 1 is not necessarily the starting riser at the bottom of the stair.

Plate 132 is a plan of a two-platform, 14 riser mitered-face stringer type stair with a continuous molded handrail starting with a volute to two quarter-turn platform turns to a landing quarter turn. The normal square tread width is to be 10", with two balusters per tread at 5" centers. Although risers 2 and 3 could be placed within the spring lines of the centerline turns so that the spacing between risers is 5" at the face of the mitered stringer, I have chosen to locate the risers at the spring lines in order to place a baluster at the platforms. The radius, therefore, will be 6-½" to the handrail centerline.

Establish the plan tangents at the volute and starting platform turn by drawing the rail centerline to meet at vertex D. Make a joint at A for the level volute rail-cap, with level tangent AB square to BD. B is then the vertex of the level and rake tangent for the lower platform section. Let the stretch out of the tangents, plus the tread at risers 3 and 4, be extended along BD and then into elevation. With the normal-rake rail height to be 2'-8" and the volute level rail-cap to be 3'-0" high, the lower tangent CD of the rake-to-rake section is pitched from vertex D to meet the centerline of the volute level tangent at vertex B. Let the rake joint for the volute section be at F, with FB the rake tangent. The layout for the rake-to-level section is the same as **Plate 76**, and the quarter-turn section is the same as at **Plate 74**.

The upper platform quarter-turn radius is the same as the lower, showing risers 12 and 13 at the spring lines. The elevation of risers 10 through 13 shows that the centerline pitches of the rail over risers 12 and 13 are the same but not in line. This difference is overcome by making both tangents equally pitched as shown, with the adjustment x in height made at the joint bevels from the center of the rail. However, block thickness must be increased to allow for this adjustment.

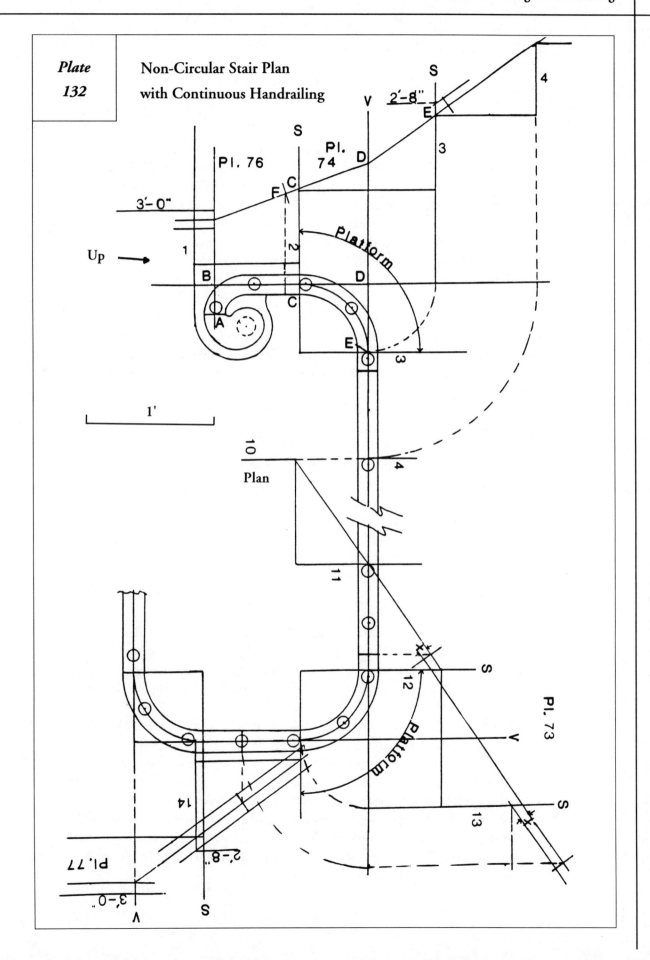

Plate 132

Non-Circular Stair Plan
with Continuous Handrailing

The location of the vertex of the landing quarter turn must now be found. Draw the centerline pitch over risers 13 and 14 as shown. Establish the 2'-8" rail height over riser 14. From the top of the floor at riser 14, set up the 3'-0" level height and rail centerline. Extend the centerline pitch over riser 14 to strike the level centerline, thereby locating the vertex. The same 6-½" radius shows that riser 14 is positioned within the turn. Consequently, the face of the top baluster is also within the spring line. All balcony balusters are then spaced 5" on centers.

Plate 133 is a starting plan of a buttress-type stringer stair instead of a mitered-face stringer. In a buttress stair, the riser positions will not affect the baluster spacing. The radius of the quarter-turn can be as desired so as to make a graceful transition throughout the turn. The stretch out elevation shows an arbitrary pitch of the lower tangent to an easement to meet the volute height of 2'-10" instead of 3'-0" shown in **Plate 132** so as to not make the ramp from C to D abrupt, or tangent joint D too high.

Plate 134 is also a buttress-type stair. Like **Plate 133**, the balusters are spaced independent of the riser positions, and the radius of the quarter turns is as desired. With the positions of risers 12, 13, and 14 established, the elevation indicates a slight easement between risers 12 and 14 if the normal rake rail height is to be maintained. However, the easement is preferably eliminated by connecting the two vertices as shown by the dotted line. This will result in a negligible increase in height over riser 13.

Plate 135 is, again, a mitered-face stringer stair plan. It is an acute angle plan of approximately 60 degrees, where the distance between risers along the tangents is equal to a normal square tread so that the pitched tangents will be equal to that over a square tread. The centerline distance between all balusters is equal. The centerline radius of the rail will be approximately 3".

Plate 136—This is an acute plan similar to **Plate 135** except it is a buttress-type stair where baluster spacing is independent of riser positions. Because of the height to be gained within such a short radius at the concave side of the handrail, it is always best to make sure that the joints are located well past the spring lines for a more graceful curve transition. **Figure 1** shows one method of pitching the tangents with a single easement at the bottom. In **Figure 2** the tangents are equally pitched with a slight easement at both sides of the turn section. While either method is satisfactory, the method in **Figure 1** is preferable since only one easement is required.

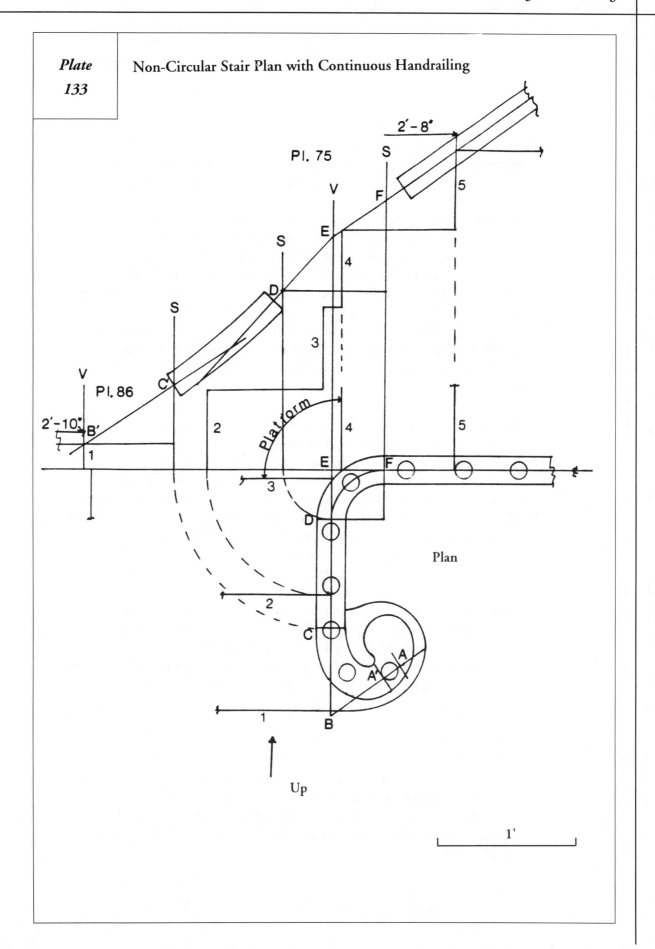

Plate 133

Non-Circular Stair Plan with Continuous Handrailing

Pl. 75

Pl. 86

2'- 8"

2'-10"

Platform

Plan

Up

1'

Plate 137 is a mitered-face stringer stair of an obtuse plan. All balusters are equal, and risers are at the spring lines. The tangents of the centerline of the obtuse turn are equal to half the width of a normal square tread. The pitched tangents will be the same as the centerline pitch over square treads.

Plate 138 is also a mitered-face stringer, single platform, obtuse plan. But, unlike **Plate 137**, where there are two balusters at a wider platform, this plan shows a narrow platform with a single baluster at the platform riser 2. The space between risers 2 and 3 is half the width of a square tread, making all balusters equally spaced. **Figures 1** and **2** show optional methods of pitching the tangents. **Figure 2** is the preferred method because, by simply beveling the straight rail joints, the easing can be made within the turn section itself. However, by doing so, the block thickness must be slightly increased. Whereas in **Figure 1**, there must be an easement outside of the lower spring line.

Plate 139 is a half-circle buttress-type stringer two-platform stair where the distance between straight run centerlines is to be 9", or 4-½" radius. With platform riser 3 established, risers 2 and 4 are to be located so that the tangent pitches and straight rail centerline pitches align. See **Plate 27**. In order for this to occur, the distance along the tangents from the spring line joint at riser 3 to the vertices, plus the centerline distance to each riser 2 and 4, must be equal to that of a square tread. Extended into elevation, the centerline pitches of the straight rail and those of the tangents are shown to align. Baluster spacing in this buttress-type stair can be as desired.

Plate 140 is a plan of a half-circle, two-platform, and mitered-face-type stringer stair with all balusters equally spaced. The centerline radius, as in **Plate 132**, is to be 6-½". Risers 2 and 4 are at the spring lines. The stretch out elevation shows the normal rail pitch to be altered at the vertices of both sections, making the layout for one section serve both. One section is simply the reverse of the other.

Plates 141, 142, and **143** are plans of winder-type stairs showing a full round wall handrail at the inside wall corner. Easements to meet normal straight rail pitches are made beyond the spring lines. Beveling straight rail joints in order to eliminate easements cannot be done with full round handrails unless the turn section is first squared as in molded-type handrails. This practice is not recommended since it is far easier to make full round easements if necessary. See **Plate 92**.

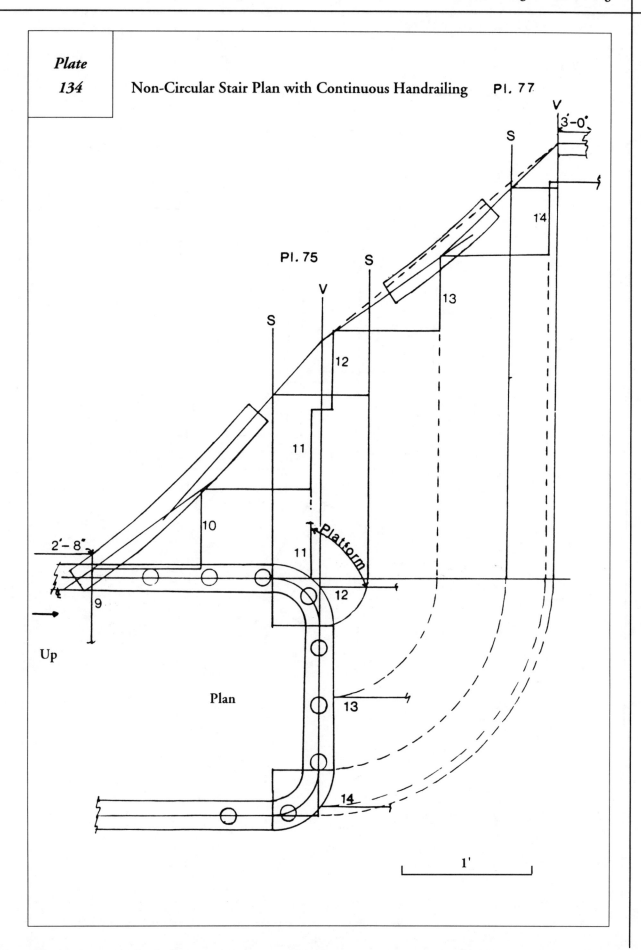

Plate 134

Non-Circular Stair Plan with Continuous Handrailing   Pl. 77

Pl. 75

Plan

Up

Platform

1'

Plate 135   Non-Circular Stair Plan with Continuous Handrailing

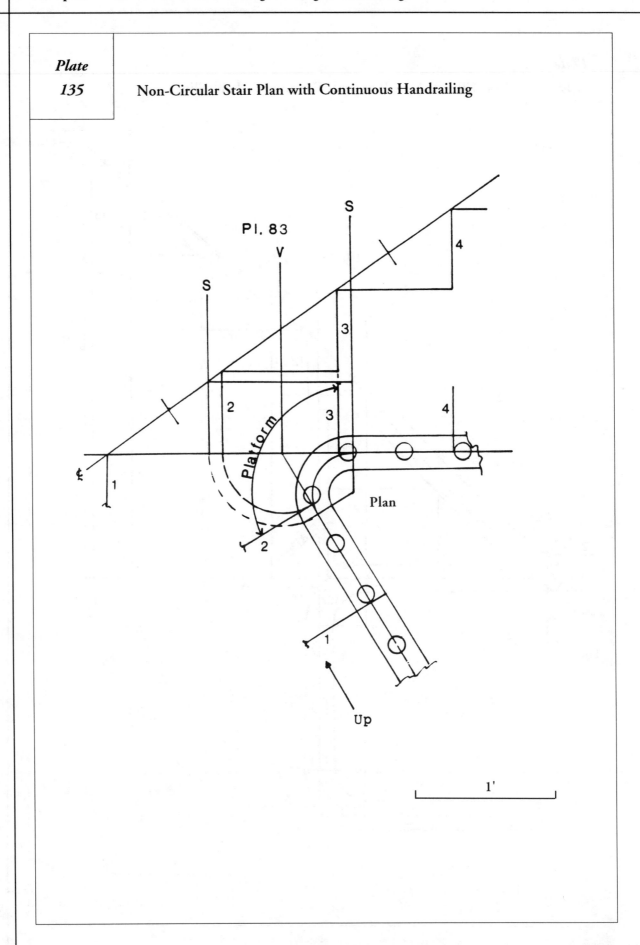

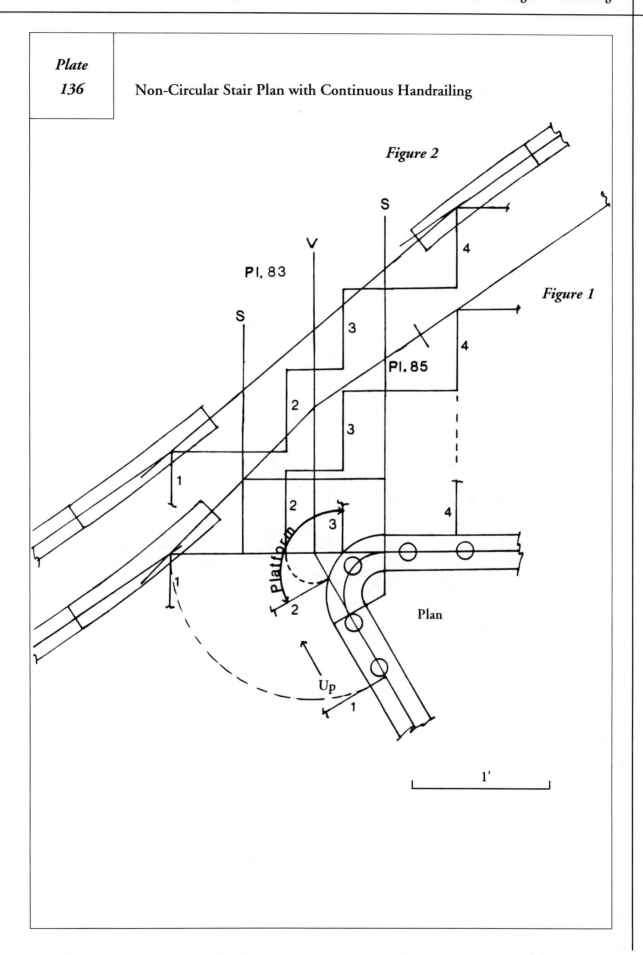

Plate 136

Non-Circular Stair Plan with Continuous Handrailing

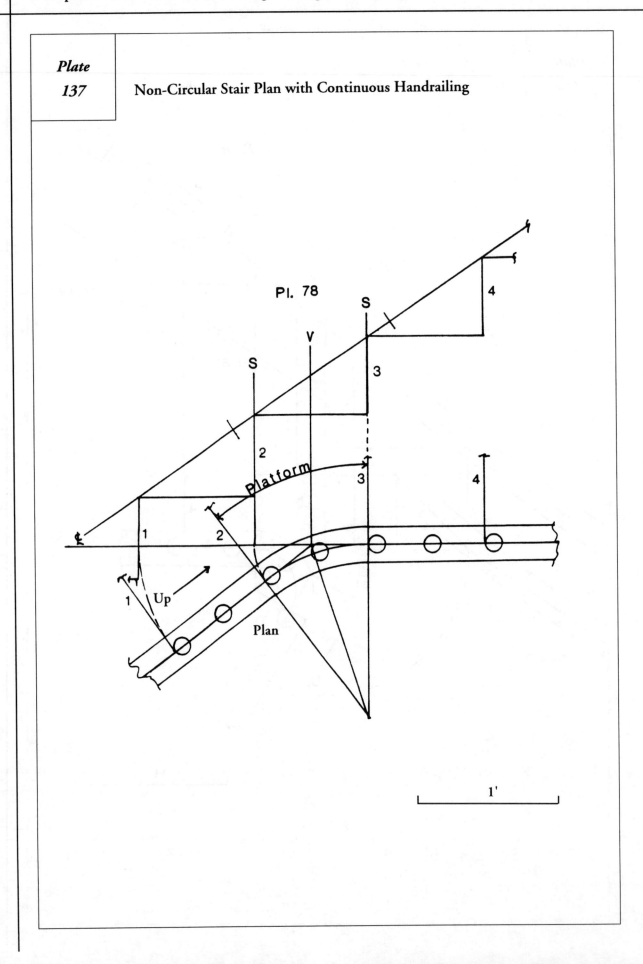

Plate 137   Non-Circular Stair Plan with Continuous Handrailing

Pl. 78

Plan

Platform

Up

1'

Plate 138

Non-Circular Stair Plan with Continuous Handrailing

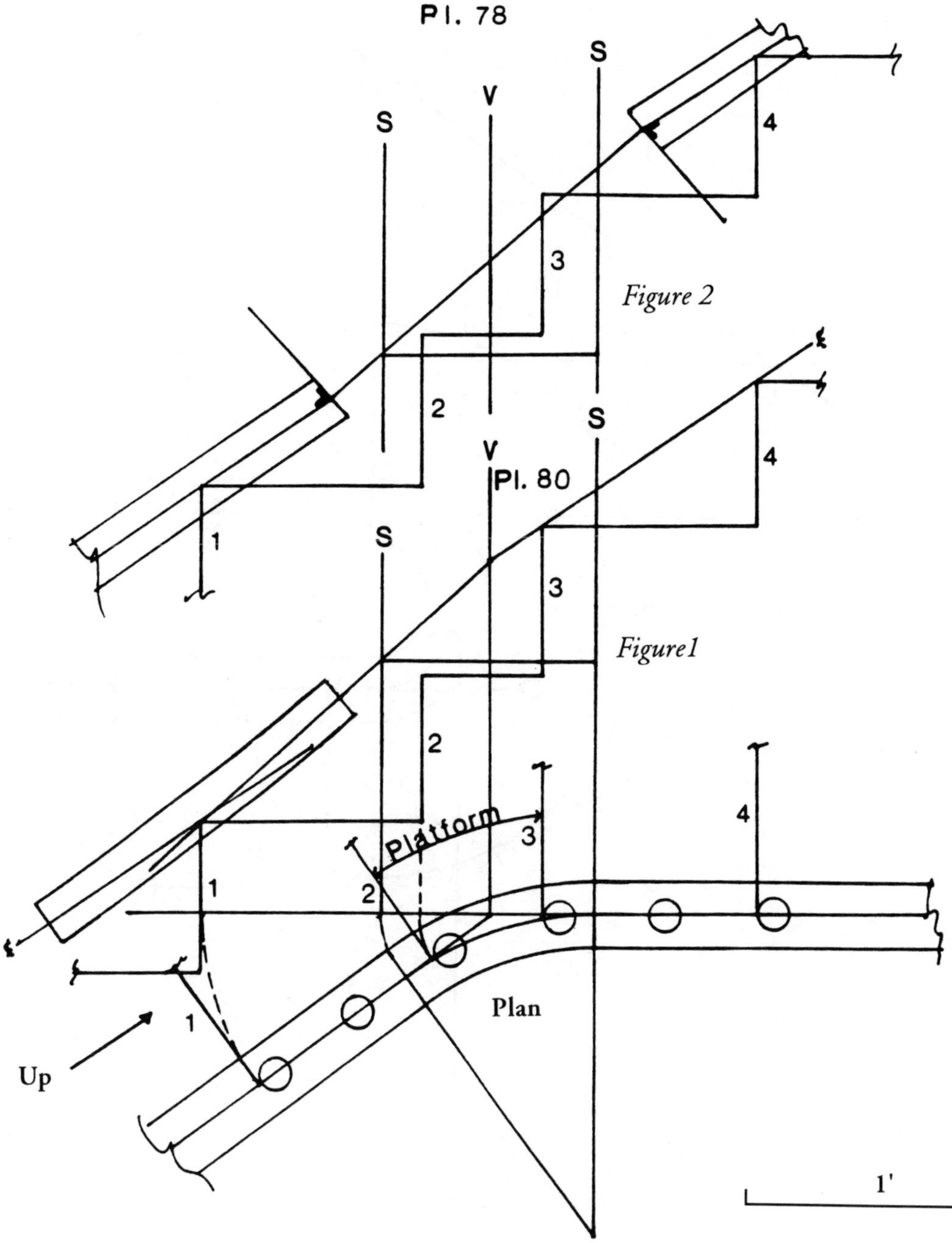

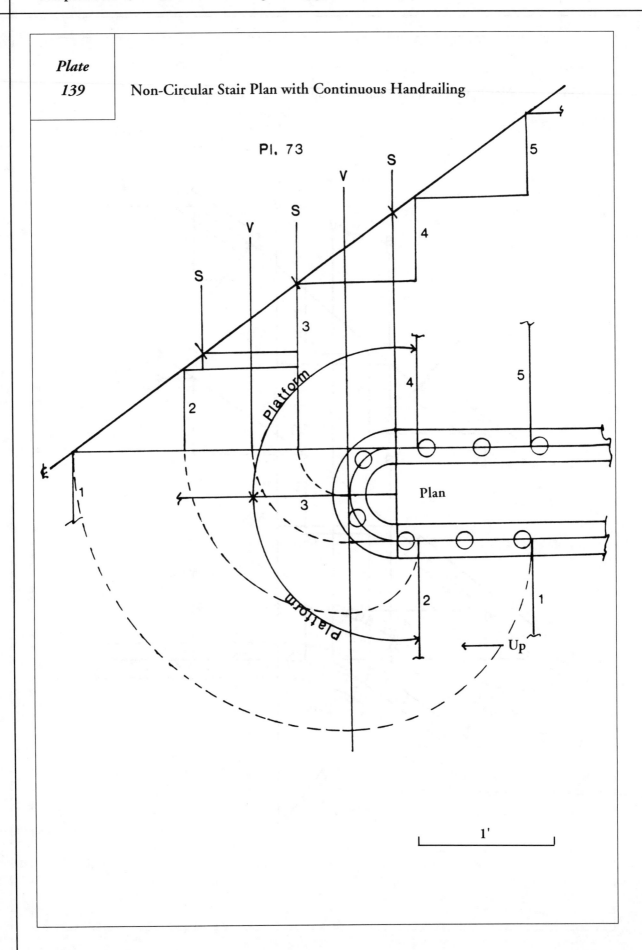

Plate 139 Non-Circular Stair Plan with Continuous Handrailing

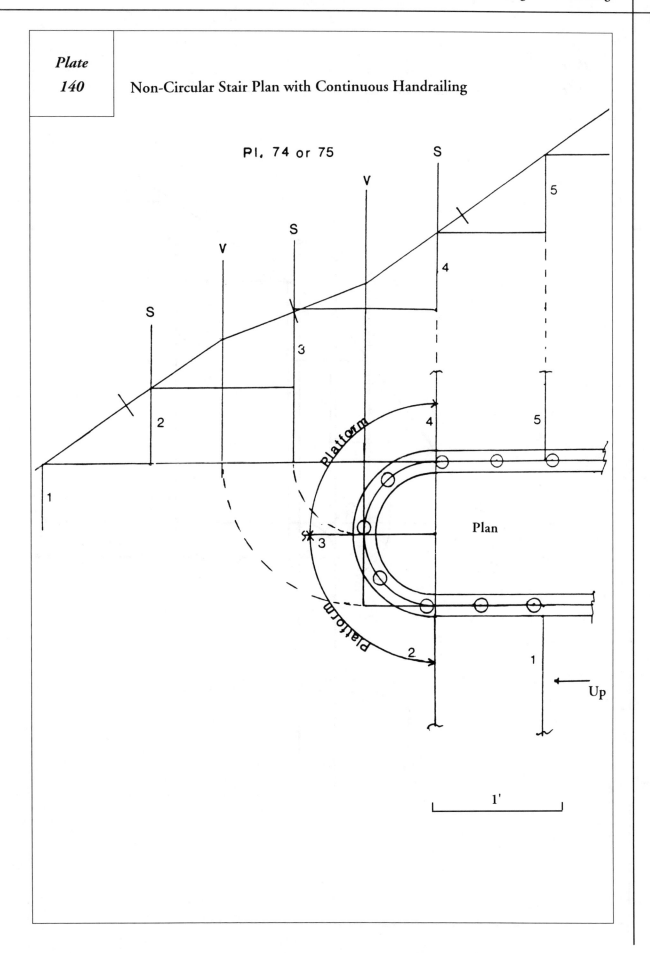

Plate 140

Non-Circular Stair Plan with Continuous Handrailing

Pl. 74 or 75

Plan

1'

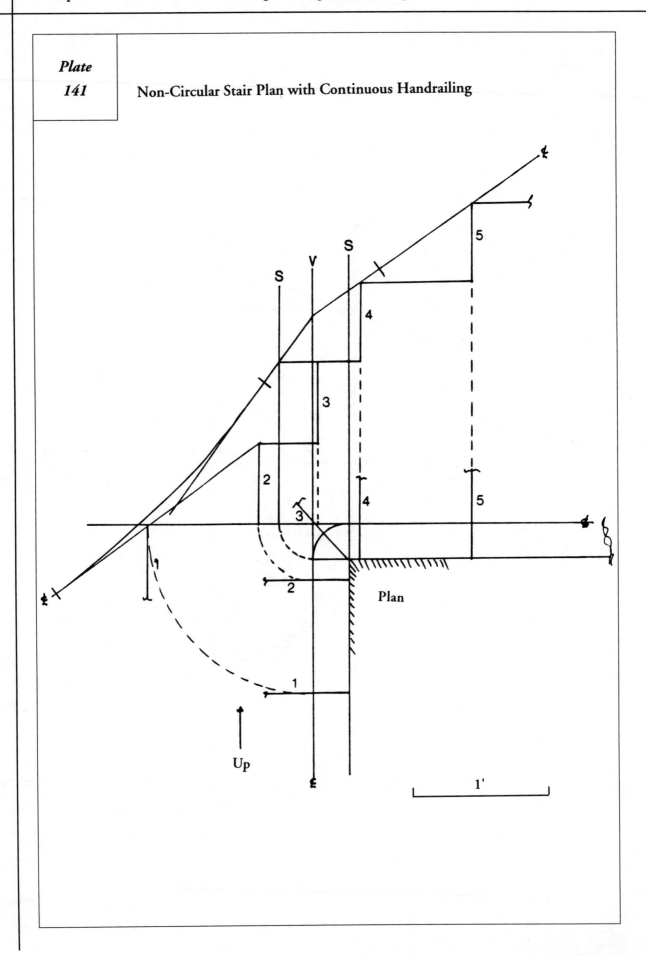

Plate
141

Non-Circular Stair Plan with Continuous Handrailing

Plan

Up

1'

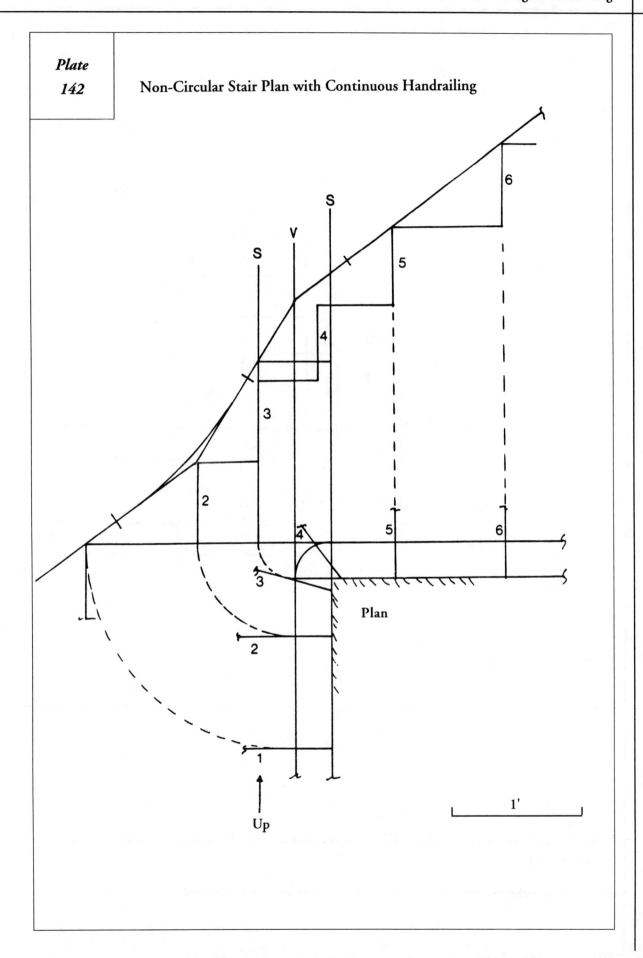

Plate
142

Non-Circular Stair Plan with Continuous Handrailing

Plan

Up

1'

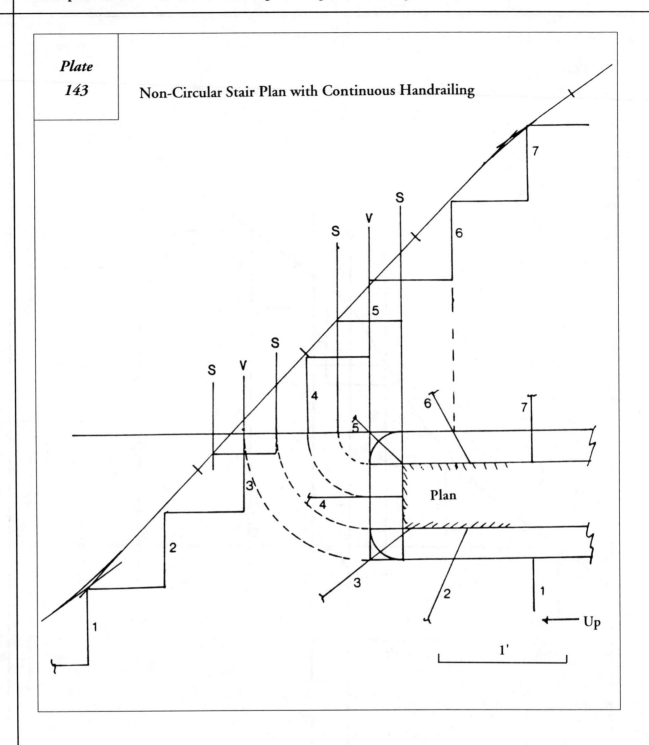

Plate 143

Non-Circular Stair Plan with Continuous Handrailing

Plan

1'

Up

Workmen bending laminated handrail to a curved form to make the handrail for a quarter-circle curved staircase.

Handrail maker shaping a right-angle incline-turn handrail with a rotary file tool.

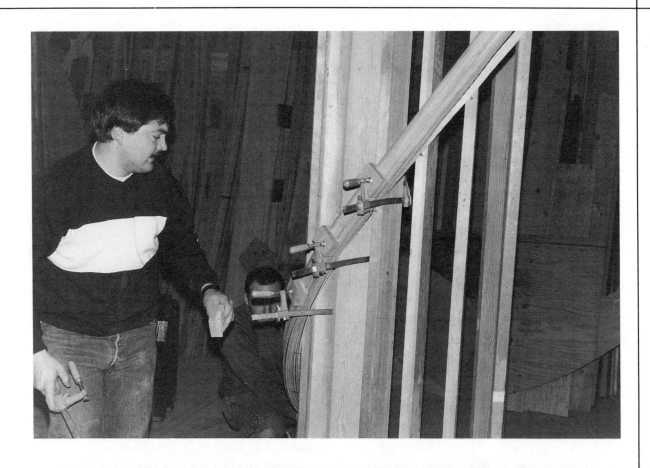

Section IV

Face Molds for Handrailing

Basic working **Plates 73** through **87** show two methods, "A" and "B", for finding the face mold and bevels required to make any incline-turn handrail section using the tangent principle.

In this section an abbreviated form of finding the face mold of **Plates 73** through **87** is shown using the method "A" principle. The previous working plates have shown how to find the angle of the tangents (**Plates 53** through **57**), the ordinate and face mold (**Plates 58** through **63**), and the bevels (**Plates 64** through **66**). The only reference shown in the following plates to those of **Plates 73** through **87** is the numbered plan and face mold parallels being equal length. Much of the line work shown in **Plates 73** through **87** has been omitted.

Each plate in this section will show the vertex point at both the plan and elevated tangents as V. The horizontal base line is marked base, and the height to be gained through the turn is indicated.

Should it be necessary to refer to the proper steps to find the face mold and bevels, refer to the following plates:

A - The ordinate and minor axis in Plates 58 through 63.

B - The angle of the elevated tangents in Plates 53 through 57.

C - The bevels in Plates 64 through 66.

D - The face mold curves in Plates 58 through 63.

Trouble Shooting for Possible Errors

Method "A"
1—Wrong Angle Elevated Tangents
Cause: Ordinate and minor axis are drawn from pitch of lower tangent.
(See Plates 58 through 63)
Cause: Distance between spring lines at the face mold is not correct. (See
Plates 53 through 57)
Cause: Line drawn from the lower spring line point of the face mold perpendicular to the pitch of the upper tangent is not properly located on the horizontal base line.

2—Wrong Bevels
Cause: Bevels taken from vertex in obtuse and acute plans. (See **Plates 65** and **66**)
Cause: Wrong dimensions to the pitched tangents taken incorrectly from where the low point of the plan tangent is squared to the horizontal base line. (See **Plates 65** and **66**)

Method "B" (See **Figure 4 Plate 64**)
1—Wrong Angle Elevated Tangents
Cause: Seat not square to ordinate.
Cause: Not proper height from seat.
Cause: HD' not marked as CK.
Cause: IB' not marked as SB.

2.—Wrong Bevels
Cause: Tangent parallels not drawn from H.
Cause: Arcs not drawn from J to tangents' parallels.
Cause: Arc points at AH not connected to K.

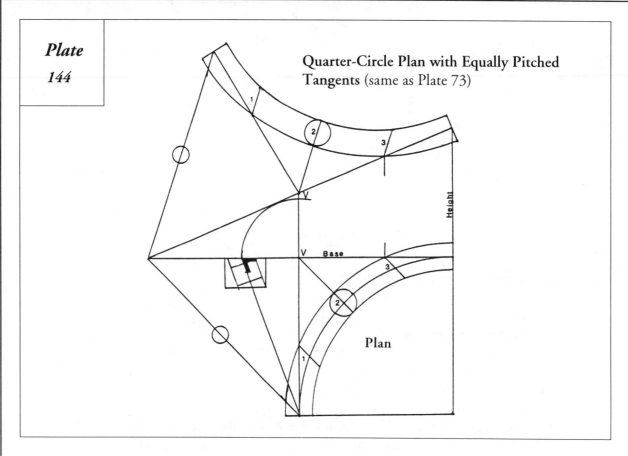

Plate 144

Quarter-Circle Plan with Equally Pitched Tangents (same as Plate 73)

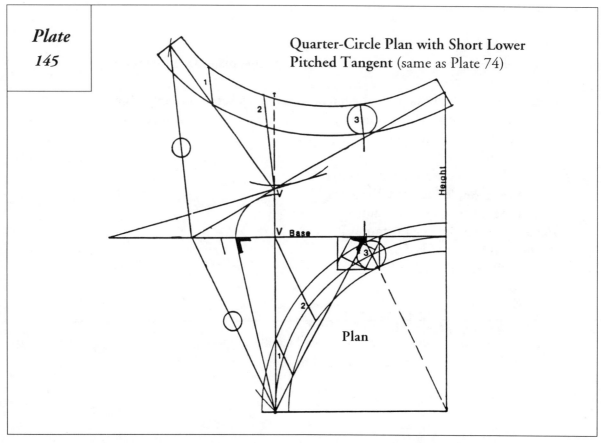

Plate 145

Quarter-Circle Plan with Short Lower Pitched Tangent (same as Plate 74)

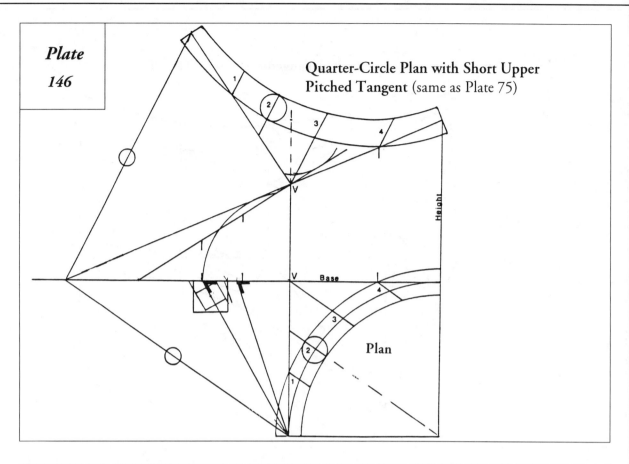

Plate 146

Quarter-Circle Plan with Short Upper Pitched Tangent (same as Plate 75)

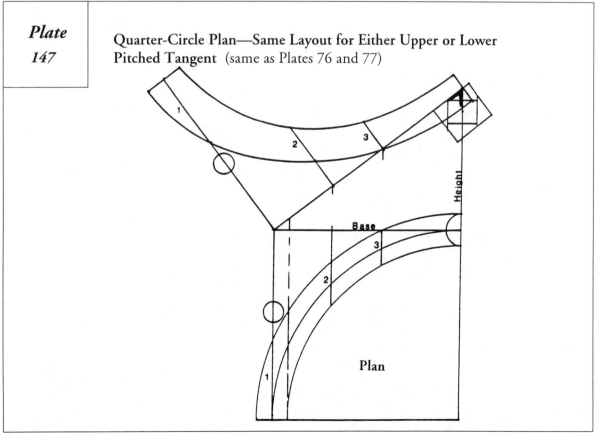

Plate 147

Quarter-Circle Plan—Same Layout for Either Upper or Lower Pitched Tangent (same as Plates 76 and 77)

Plate 148

Obtuse Plan—Equally Pitched Tangents (same as Plate 78)

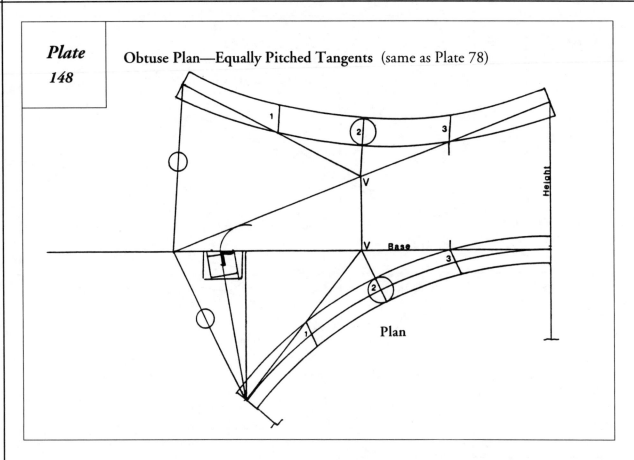

Plan

Plate 149

Obtuse Plan—Short Lower-Pitched Tangent (same as Plate 79)

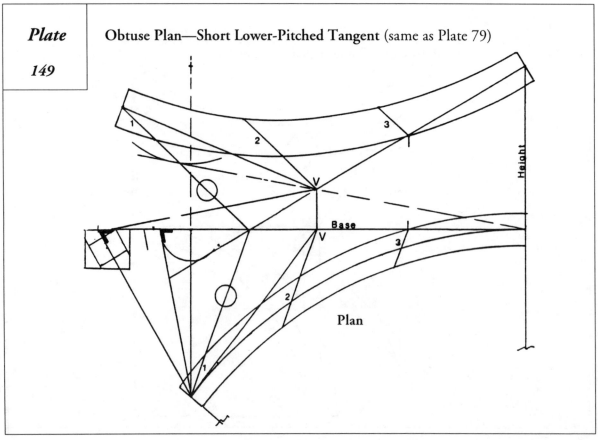

Plan

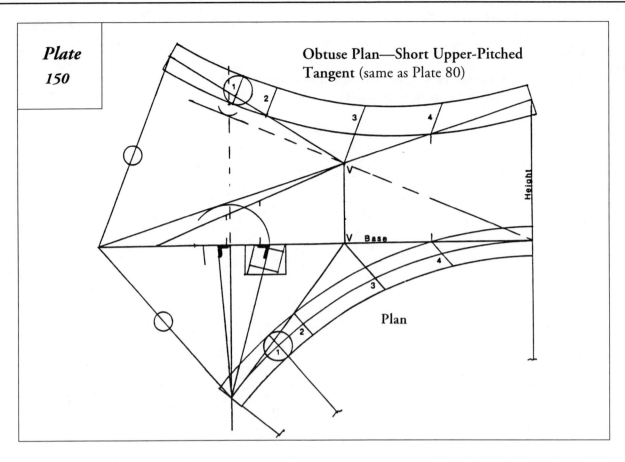

Plate 150

Obtuse Plan—Short Upper-Pitched
Tangent (same as Plate 80)

Plan

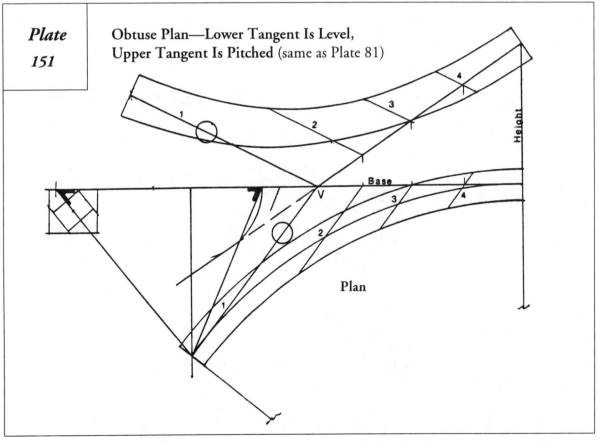

Plate 151

Obtuse Plan—Lower Tangent Is Level,
Upper Tangent Is Pitched (same as Plate 81)

Plan

Plate

152

Obtuse Plan—Upper Tangent Is Level,
Lower Tangent Is Pitched (same as Plate 82)

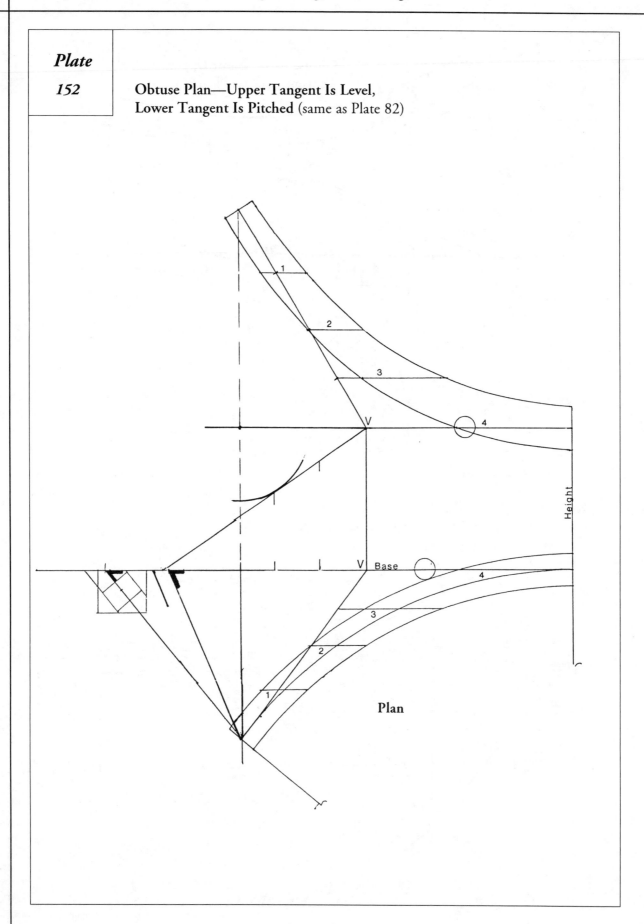

Plan

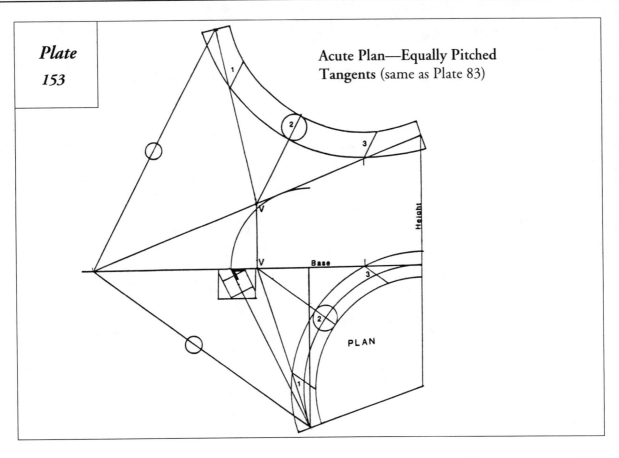

Plate 153

Acute Plan—Equally Pitched Tangents (same as Plate 83)

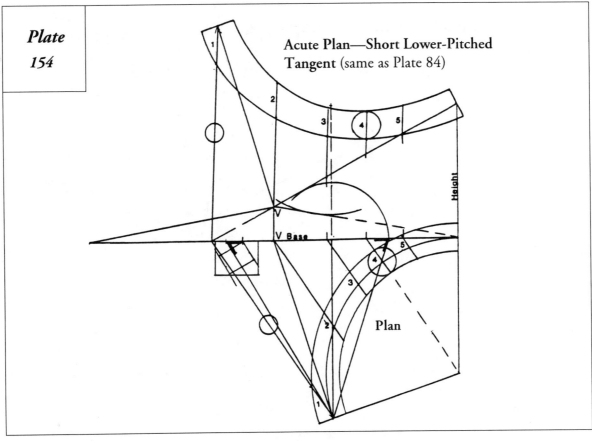

Plate 154

Acute Plan—Short Lower-Pitched Tangent (same as Plate 84)

Plate 155

Acute Plan—Short Upper-Pitched Tangent (same as Plate 85)

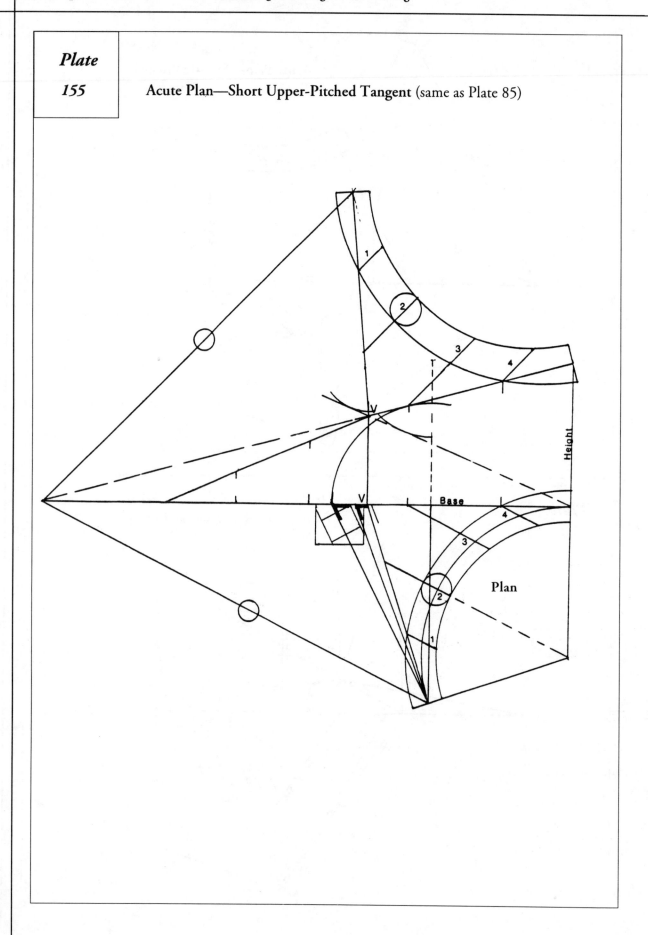

Plate 156

Acute Plan—Upper Tangent Is Pitched, Lower Tangent Is Level
(same as Plate 86)

Plate 157

Acute Plan—Lower Tangent Is Pitched, Upper Tangent Is Level (same as Plate 87)

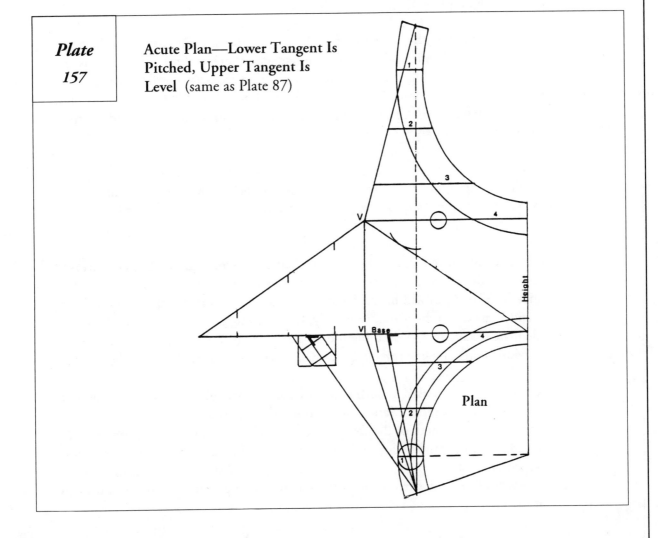

Section V

Making Tangent Handrail Sections To Fit on Top of an Existing Metal Railing

One of the most time consuming tasks in handrailing is fitting handrail blocks to the top of an existing incline-turn metal subrailing by whittling the blocks to suit the metal through a trial-and-error process at the job site. The blocks must not only fit on top of the metal but are normally plowed to receive the metal. Once fitted, joints for adjoining sections are made and the blocks returned to the shop for squaring, joining, shaping, and sanding.

A more efficient method of making such rail sections is to use the tangent system already described and detailed in foregoing plates, thereby eliminating tedious and time consuming fitting at the job site. However, in order to use the tangent method, dimensions must be taken in order to draw a floor plan of the turn condition. The height through the turn and the curved ramp of the convex side of the curved metal must also be taken.

Plate 158, Figure 1, is a side view of an incline-turn metal subrailing of a "U" turn plan inclining from left to right with straight rail below and above the turn. Mark the lower and upper spring line points on the outside curve of the metal as A and B.

Below A and above B are straight rail pitched as shown. A method I have successfully used to take field dimensions is to take a wood straight edge, say 3/8" to 3/4" thick by 3" or 4" in width, and lay it across the metal with its edge plumb over the A and B marks on the metal. Normally, several straight edges of 2', 4', and 6' are adequate to take most field dimensions. The pitched straight edge must lie level to its width along any line (preferably a shallow saw cut) squared to the edge.

A leveling device is used at both ends to secure the straight edge to the metal in its level-to-width position. Mark numbered lines, 2" apart and square to the edge, along the entire length of the straight edge. Level dimensions are then taken from the edge of the straight edge to the top of the convex side of the metal curve with a carpenter's steel square. The body of the square will rest along the width of any numbered line, with the tongue of the square falling plumb against the convex side of the metal.

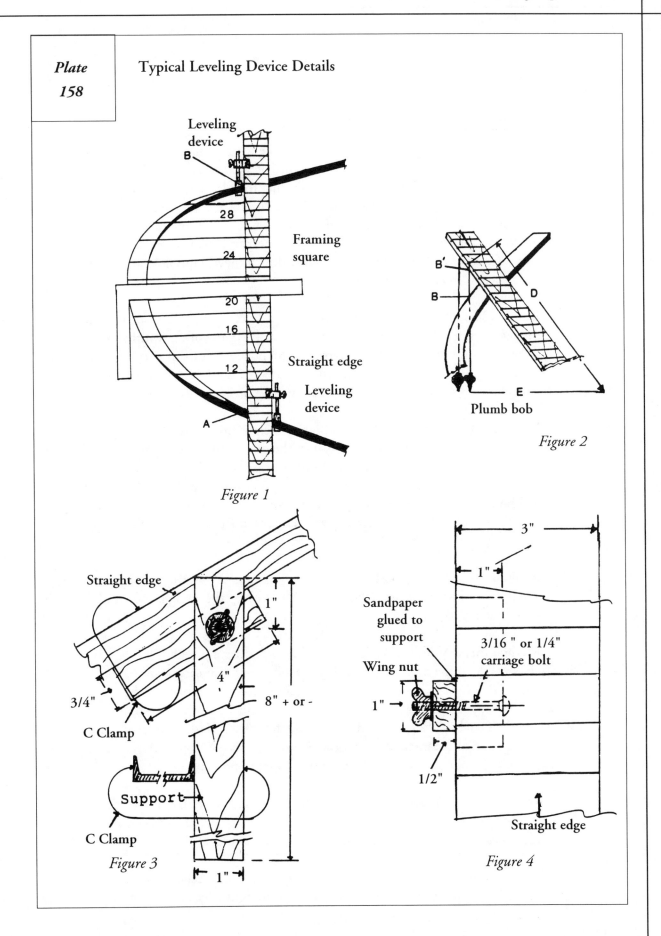

Plate 158

Typical Leveling Device Details

Leveling device

B

28

24

Framing square

20

16

Straight edge

12

Leveling device

A

Figure 1

B'

B

D

E

Plumb bob

Figure 2

Straight edge

1"

4"

3/4"

8" + or -

C Clamp

Support

C Clamp

1"

Figure 3

3"

1"

Sandpaper glued to support

3/16 " or 1/4" carriage bolt

Wing nut

1"

1/2"

Straight edge

Figure 4

Mark down the level dimension and line number for each dimension taken from the straight edge to the convex side of the metal. The more level dimensions that are taken, the greater the assurance of an accurate plan. However, for purposes of showing the method of finding the plan curves, let us take only the level dimensions along lines 12, 16, 20, 24, and 28. Mark the metal at each ordinate dimension. From these marked points on the metal, draw a line across the width of the metal reasonably perpendicular to the convex curve. Check across these perpendicular widths for level, and note any height difference from one edge to the other.

If the widths show less than 1/8" difference in level, regardless of the width of the metal, make the plow at the rail bottom 1/4". If any of the widths show 1/8" or more out of level, then the plow should be 3/8". For this particular turn section, let us assume the metal is less than 1/8" out of level across any point. Therefore a 1/4" plow at the rail bottom will be proper.

Figure 2—Of three methods of finding the height from A to B, two should be found as a means to check one against the other. All three use the plumb bob. The following procedures will be further explained in **Plate 159.**

Method one: Drop a plumb line from the top of the straight edge along point B at the metal. Take the shortest (level) dimension from A to the plumb line cord as E.

Method two: Drop a plumb line from the top end of the straight edge. With a long-bladed bevel square, take the bevel formed by the straight edge and the plumb line cord.

Method three: With the straight edge removed, drop a plumb line from the convex side of the metal at point B. Take a piece of thin cardboard or tagboard at least 10" wide and long enough to make a template of the convex side of the metal ramp between A and B, and slide it between the plumb line and the metal. Mark the plumb line on the template. Mark the top of the convex side of the metal ramp on the template at least 6" below A and above B along the pitch of the metal, being sure to mark points A and B.

As one of the two methods to find the height from A to B, try to include method one if possible. Be certain to take the plumb dimension from B', at the bottom of the straight edge, to the top of the metal at B.

Figures 3 and 4 show details of the leveling device, of which two are required. Small C clamps fasten the device to both the straight edge and the metal.

Plate 159

Finding Plumb Height

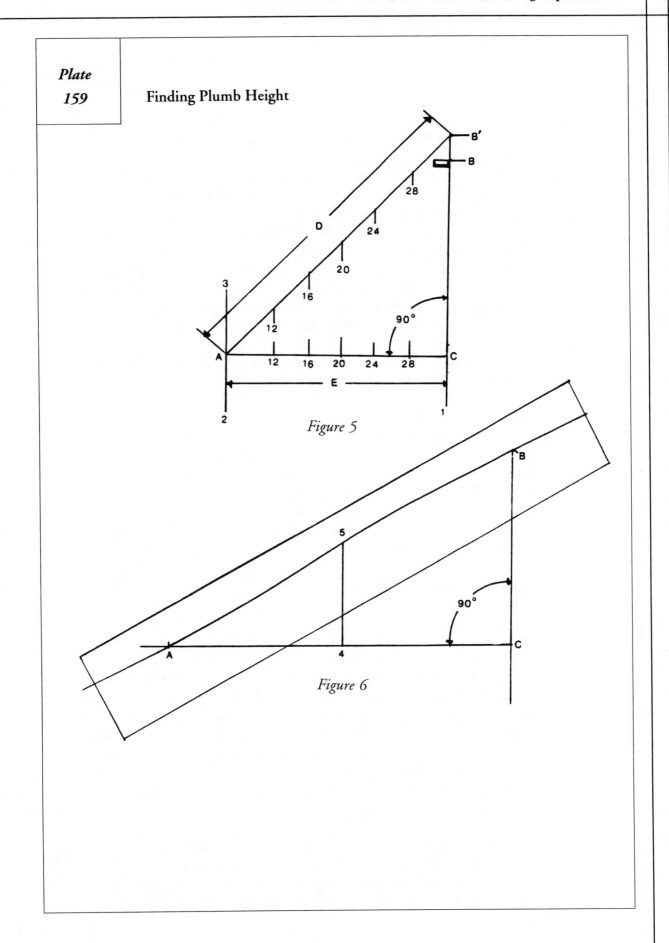

Figure 5

Figure 6

Figure 5, Plate 159, shows the method of finding the plumb height from A to B using method one of **Figure 2**. Draw a vertical line, such as B'1. Mark the level dimension E taken from A to the plumb bob line shown in **Figure 2** and draw parallel line 2-3. From B', mark the length D, shown in **Figure 2**, to intersect parallel line 2-3 at A. Make AC perpendicular to B'1. From B', plumb down the dimension B'B taken in **Figure 2** from the bottom of the straight edge to the top of the metal at B. BC is then the plumb height through the turn from A to B. Angle CB'A should be the same as the bevel taken along the pitched straight edge and the plumb-line cord, as method two in **Figure 2**. To prove method two, with the bevel and length D known factors, the perpendicular line AC is drawn from B'1 to establish point C in Figure 5. This establishes the height from A to B as BC. Along AB', mark straight edge lines 12 through 28 taken from straight edge length D. Transfer 12 through 28 from AB' plumb to the level line E.

Figure 6 shows the template stretched out on a flat surface. The vertical line drawn from B is an extension of the plumb line marked on the template in method three of Figure 2. Draw a perpendicular line from the plumb line to intersect point A on the template. Mark the right angle point on the plumb line as C. CB in **Figure 5** should equal CB in this figure. AC is the stretch-out curve of the convex side of the metal at the plan. Since the handrail for the entire curve of the metal is to be made in two sections, draw perpendicular 4-5 at the center of AC.

Figure 7—Draw A'C', with marks 12 through 28, equal to length E in Figure 5. Extend 12 through 28 vertically from A' to C' to equal the level ordinate lengths taken in Figure 1, Plate 158. Mark length AC, in Figure 6, on a thin lath. Bend the lath with marks A and C fixed on A' and C'. Assuming width A'C' to be correct, the bent lath should touch the end of all vertical extensions of 12 through 28, especially the point of maximum width. If all of the extended points cannot be touched, then ordinate lengths are not accurate. Therefore, draw the curve as nearly as possible to all points without altering points AC on the lath or A'C' on the plan. To overcome the problem of a slightly inaccurate plan curve, when the bottom of the rail is plowed, the plow should be made 3/8" deep and approximately 1/4" less than the metal width. The exact plow width to suit the metal can then be done when other joints are made at the job site.

In this particular plan, the curves of the metal can be drawn from a radius center point at O. Therefore, strike all curves of the metal and rail width from O. Draw random lines a through e extended to the rail width. Make FG parallel to A'C'. Let JH, the stretch out of the convex side of the metal

Plate 160

Finding Plumb Height and Stretch-Out Ramp

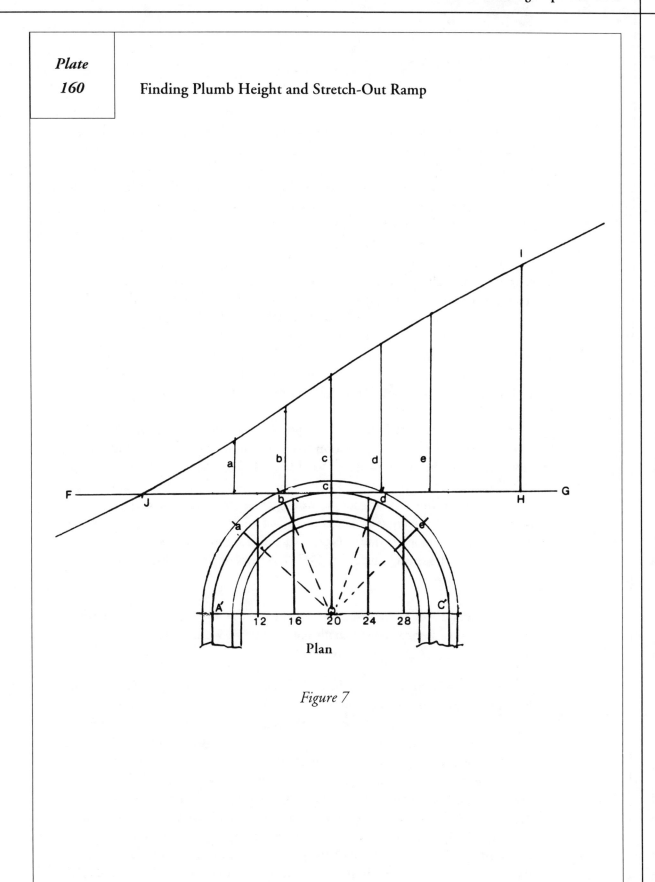

Plan

Figure 7

from A' to C' include points <u>a</u> through <u>e</u>. From H, draw HI perpendicular to JH and equal to CB in **Figures 5** and **6**. Now, cut the cardboard template along the plumb line and the marked ramp. Place point B of the template on point I, with the plumb line falling along IH. Point A of the template should then fall on point J. The ramp is then traced from J to I. Extend stretch out points <u>a</u> through <u>e</u> vertically to intersect the drawn ramp.

Figure 8, in **Plate 161**, shows the same plan as in **Figure 7, Plate 160**. Only the centerline and rail width are shown. The "U" turn is to be made up of two equal rail sections. Plan tangents are shown as QR, RS, ST, and TU. The stretch out of the tangents is shown along RS and ST. Height U'I' is the same as BC in **Figures 5** and **6, Plate 159**. The pitches below Q' and above I' are the same and are extended to strike the vertex lines at R' and T'. Tangents connecting R' and T' show that SS' is also the same height as 4-5 in **Figure 6, Plate 159**, as it should be, since S' is the center of the joint of both sections.

Since the tangents of both sections are pitched the same, one section being the reverse of the other, only one section has to be laid out in order to find the face mold and bevels for both sections. See **Plate 74**. The lesser bevel for joint S' is at Y. The greater bevel for the joint at V' or I' is at X. Draw the rectangular size of the rail as shown, allowing ample wood above and below the diagonal corners of the block thickness Z. ZZ is the block width so that the face mold can be drawn on the block surfaces. Note that the face mold shows some straight wood beyond the spring line at V'. Only the sides of the sections are to be dressed at this time.

Figures 9 and 10 show the stretch out ramps of the top of the metal at both the inside and outside curves of the handrail. In **Figure 9** the stretch out of the inside curve, 6 to 7 in **Figure 8**, is shown as KL.

In **Figure 10** the stretch out of the outside curve, 8 to 9 in **Figure 8**, is shown at PP'. The thin strips taken in **Figure 7** of both inside and outside curves of **Figures 9** and **10**, should also include the radius line marks <u>a</u> through <u>e</u> where they strike the curves. The stretch out lines of both **Figures 9** and **10** show the positions of <u>a</u> through <u>e</u>. In both figures, <u>a</u> through <u>e</u> are extended vertically the same heights as they are in **Figure 7, Plate 160**. The centerline pitch below K of **Figure 9** and P of **Figure 10** is the same as the centerline pitch below Q in **Figure 8**. Likewise, the centerline pitch above N of **Figure 9** and M of **Figure 10** is the same as the centerline pitch above I in **Figure 8**. In **Figure 9**, to find the stretch out ramp at the top of the metal for the inside, or concave side, of the assembled handrail sections, simply draw the ramp from K to N touching the heights <u>a</u> through <u>e</u>. The same procedure is followed to find the stretch out

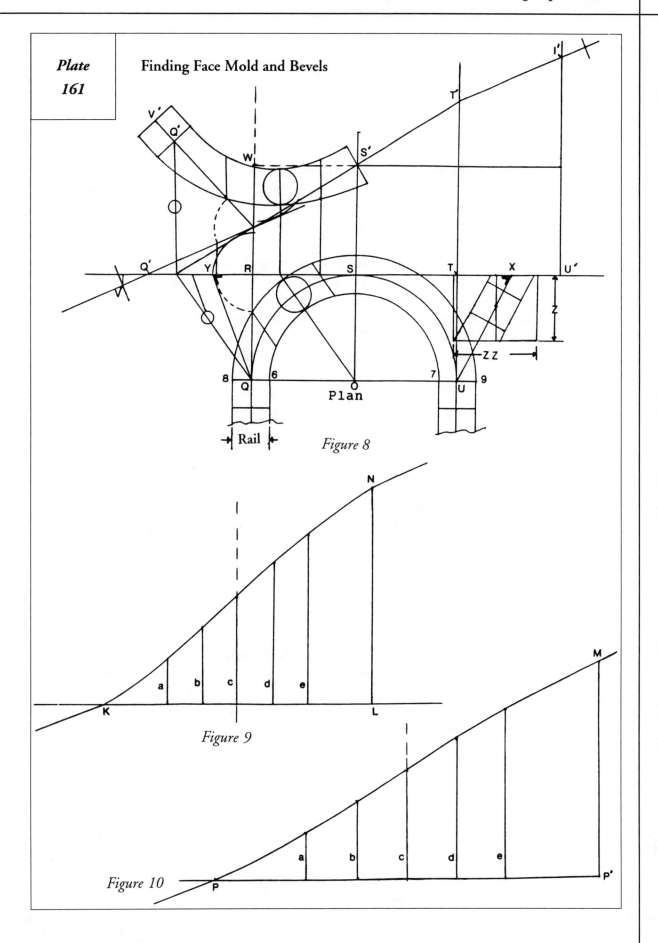

Plate 161

Finding Face Mold and Bevels

Figure 8

Figure 9

Figure 10

ramp in **Figure 10.** Cut templates of thin cardboard of the two ramps.

Figure 11, Plate 162—With the sides of the rail sections dressed plumb, insert a short metal pin of a 6d or 8d nail into the center of joint S' (**Figure 8**) of either section. Join the two sections with "dogs", as shown in this perspective. Mark K, N, M, and P 1/4" above the normal rail bottom along the plumbed spring lines. With N of the template NK (**Figure 9**) placed on mark N of the rail block, trace the inside ramp shown by the dotted curve. Scribe 1/4" below the dotted curve for the rail bottom. Mark the rail bottom on the outside curve in the same manner using the outside template. The sections can now be taken apart and the pin removed. The bottom is roughly dressed to the mark shown. The two sections can now be bolted together with the lower section receiving the lag end of the bolt. The sections can now be dressed to the marked line. The plow to receive the metal is made 1/4" deep and 1/16" wider than the metal. If there is a slight difference between the stretch outs of **Plate 159** and **Plate 160**, then make the plow 3/8" deep. Do not dress the rail top at this time.

The sections are now taken to the job site for finish fitting over the metal. The finished rail bottom should be about 1/4" below the top of the metal. Mark the desired rail bottom if necessary, dress the joints to receive the straight rail, and return the rail to the shop for squaring, shaping, and sanding to complete the job. The completed rail can now be returned to the job site and secured to the metal.

Figure 12 is a perspective of the twisted handrail section plowed to fit over the metal.

Although the railing, in this instance, is made to fit a "U" turn condition, the same method is used to make the handrail for any incline-turn metal condition. In marking the plan tangents for any curve, if a radius point cannot be found to draw the curve, make the tangents right angle to a line drawn as near to perpendicular to the curve as can be determined.

Shaping the Handrail

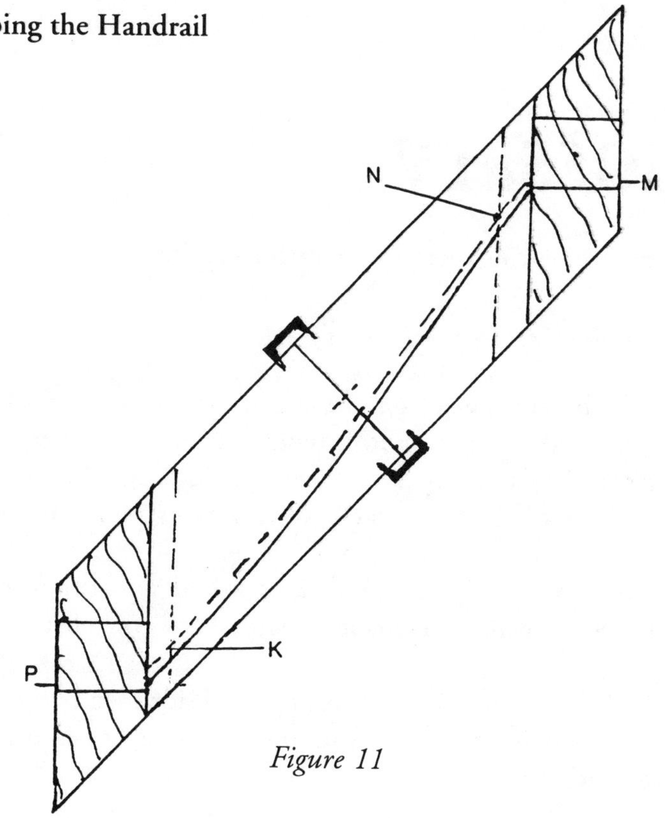

Figure 11

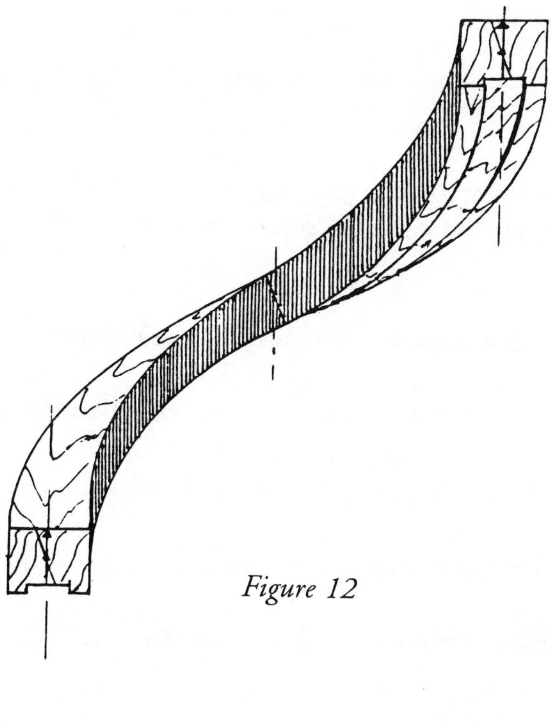

Figure 12

Glossary

Baluster - Supporting member beneath a handrail.

Base - A hollow, rectangular wood receptacle for a handrail post.

Bevel - The bevel applied from the surface of a handrail block through the center of the rail at a butt joint. It represents the plan tangent being projected across the face of the joint when the block stands in its position over the plan, the entire length of the bevel line being plumb to the plan tangent.

Box Corner - Vertical stringer members around a partition corner joining straight-run stringers, generally in a switch-back type of stair.

Bracket - 1. A return right angle ornamental side member to the riser at the open side of a stair stringer. 2. A supporting member beneath the tread, as in a timber bracket.

Bull-Nosed Tread - Starting stair tread rounded to width at one or both ends.

Buttress - Built-up width of a stair stringer or floor piece to receive balusters in a shoe.

Carriage - A type of stringer or supporting member for the treads. It is generally made from 2"x12" stock and cut out to the riser height and tread width.

Cove Molding - The molding beneath the nosing and against the riser.

Easement - The over or under ramp of a stair stringer or a handrail.

Elevated Tangents - The pitch of unfolded plan tangents.

Elevated Tangents' Angle - The angle the tangents form as they unfold.

Ellipse - Non-radius oval drawn by the string-and-pin, trammel, or ordinate method.

Face Mold - Elliptical pattern used in forming the side curves of the handrail block for a climbing-turn handrail section. Its shape is that of the inside and outside rail-width plan curves projected to the surface of the oblique plane parallelogram formed by the elevated tangents and their respective parallels. Tangent lines to guide its application to the handrail block are marked on its surface.

Fascia - A flat trim member under a balcony landing strip, or a side member of a stringer buttress.

Furring - Built-up material between finish members as in a stair stringer or floor-level buttress.

Glue Block - Generally used between the back of a riser and a stair tread or any hidden area to secure two right angle members together.

Header - In stairbuilding it is the level balcony rough framing area surrounding a stairwell.

Height - In handrailing it is the center-to-center vertical distance an incline-turn handrail section must gain through the turn. In stairbuilding it is the total height from one floor to the next.

Horizontal Base Line - In handrailing it is a level line originating from where the pitch of the lower tangent intersects the vertical extension of the lower spring line to where it meets the vertical extension of the upper spring line.

Housing - The recessed routing of a stair stringer to receive treads and risers.

Incline-Turn - A pitched and side-curved handrail section.

Kerf - A saw cut across the width of a board to aid in bending.

Keying - A thin, solid wood member glued into the kerf of a curved stringer or fascia.

King Post - The starting post beneath a volute or turnout rail cap.

Laminating - In both stairbuilding and handrailing it is the gluing of thin members together to make a specific thickness usually for a curved stair stringer or handrail.

Landing - The platform or top floor level of a stair.

Landing Strip - Top tread floor piece or floor piece at a balcony level.

Layout - Full size drawing of the stair plan, or geometrical drawing to find the face mold for a handrail section.

Major Axis - The long diameter of an ellipse. Its use in handrailing is to find the shape of the face mold using either the trammel or string-and-pin method.

Minor Axis - The short diameter of an ellipse. Like the major axis it is used to find the face mold using either the trammel or string-and-pin method. However, it can also be used to find the face mold using the ordinate method. The semi-minor axis is the plan ordinate projected to the oblique plane.

Mitered Face Stringer - Supporting member under the stair treads and mitered to the risers at the open side of a stair.

Newel - Starting or landing post of a staircase.

Oblique Plane - The inclined parallelogram formed by the elevated tangents angle and their respective parallels. Also termed an oblique parallelogram.

Ordinate - In handrailing the ordinate is the guiding instrument for finding the necessary face mold pattern in order to make an incline-turn handrail section. Its plan direction is such that when projected to the oblique plane it becomes the line and direction of the minor axis, from which the face mold can be found.

Ordinate Direction - The directing level line to which the ordinate is drawn parallel from the radius center. It begins from where the pitch of the upper tangent strikes the horizontal base line, which point is transferred to the extended plan line of the upper tangent, and then drawn to meet the low point of the lower plan tangent.

Parallelogram - Four-sided geometrical figure with opposite sides parallel.

Parallels - Lines equidistant apart throughout their lengths.

Pitch - An incline, as opposed to horizontal or vertical.

Pitch Board - A right triangle thin pattern with one leg the width of a tread and the other leg the height of a riser.

Plan - A floor view of the stair or handrail section drawn either to scale or full size.

Plank Pitch - Pitch of the handrail block perpendicular to the ordinate or ordinate direction from lower to upper spring lines.

Platform - The landing between runs of stairs.

Projection - The amount the tread protrudes beyond the riser face.

Quarter Turn - A ninety degree or right angle turn.

Rail Block - The block from which a climbing-turn handrail section is made.

Rail Section - Rail section formed by the lower and upper plan tangents and the angle of the plan tangents.

Return Nosing - Separate nosing piece mitered at the end of a tread to the tread nosing, and equal in width to the nosing projection of the tread from the face of the riser.

Riser - The vertical height from the top of one tread to another. The solid member from the bottom of a top tread to the bottom of a lower tread.

Riser, Deep - Vertical riser from the bottom of the nosing of a tread to the bottom of a stringer.

Riser, Kick-Back - Slanted riser from the front of a tread nosing to the slanted back edge of the lower tread.

Run - The level width of a tread between the face of two risers of a staircase. Also called the "cut". A series of treads and risers in a stringer.

Scroll - Same as a volute, handrail's spiral starting at staircase's bottom.

Seat - In handrailing it is the level distance from the lower plan spring line to the upper plan spring line perpendicular to the ordinate direction.

Semi-Circle - Half of a full circle. A tangent handrail section should always fall within the semi-circle.

Semi-Ellipse - Half of an ellipse. The face mold for a handrail section must fall within the semi-ellipse.

Shoe - The bottom member upon which a baluster may rest.

Soffit - The underneath exposed portion of a staircase. The ceiling along the balcony beneath the stair to which a soffit mold is to be applied.

Spring Line - The point where a tangent line touches the curve square to a radius line.

Squared - Rectangularity of the handrail prior to molding to profile.

Square Tread - A normal straight tread in a stair.

Stairwell - The area in which the stair is located.

Stave - A solid wood member much wider than a keyed member, curved at the bottom, and glued to a curved veneer over a form. It is not inserted into a kerf as is the keyed member.

Story Rod - In stairbuilding it is a floor to floor rod with the risers of the stair marked.

Stretch Out - Unfolding of the plan tangents along a straight line. Also unfolding the riser positions on the tangents along a straight line. The length of a curved line along a straight line.

Stringer - The supporting member of the treads and risers of a stair.

Switch-Back Stair - Two opposite straight runs of stairs within the same staircase.

T-Gauge - A device used with a pitch board to mark the tread and riser at an established margin from either the top or bottom edge of a stringer.

Tangent - A line touching, but not crossing a curved line. It is perpendicular to the radius of the curve. It is also as perpendicular to an elliptical curved line as can be determined.

Timber - A supporting member under the treads and risers of a staircase separate from the stringers.

Tread - The horizontal walking member between two risers of a staircase.

Trammel - An instrument for drawing an ellipse.

Turnout - A side-turned starting handrail section to a rail cap.

Twist - The helical turn of a climbing-turn handrail section.

Unfolded Tangents - Plan tangents of the handrail centerline stretched out along a straight line.

Veneer - A thin member.

Vertex - The meeting point of two tangents forming an angle.

Vertex Line - A vertical line of the vertex.

Volute - Same as a scroll, the spiral starting of the handrail at the bottom of a staircase.

Wedge - The glued and nailed tightening piece beneath the tread and against the riser in a housed-and-wedged type stringer.

Winder Tread - A tread that is wide at one end and narrow at the other.

This compact quarter-circle stair is made of clear red oak. Wall stringers are 1" thick and housed-and-wedged. Open-side 1" thick laminated stringers are mitered to 3/8" thick sawn brackets at the risers. The 2-1/4" thick by 2-1/2" wide tangent-made handrail features the wood grain as it gracefully eases from the level volute caps to terminate at the wall.

Index

Managing Editor	Richard Sorsky
Illustrator	George R. di Cristina
Copy Editor	Kari Vaughan
Book Design	Matt Hayden
Page Production	The Red Dog Press
Indexer	Harriet Hodges
Typeface	Adobe Garamond
Paper	Publishers 60 lb
Printer	McNaughton & Gunn, Saline, Michigan